LEADED GLASS

a handbook of techniques

To Peter and Joanne Nervo: two luminaries in the world of glass lampshades

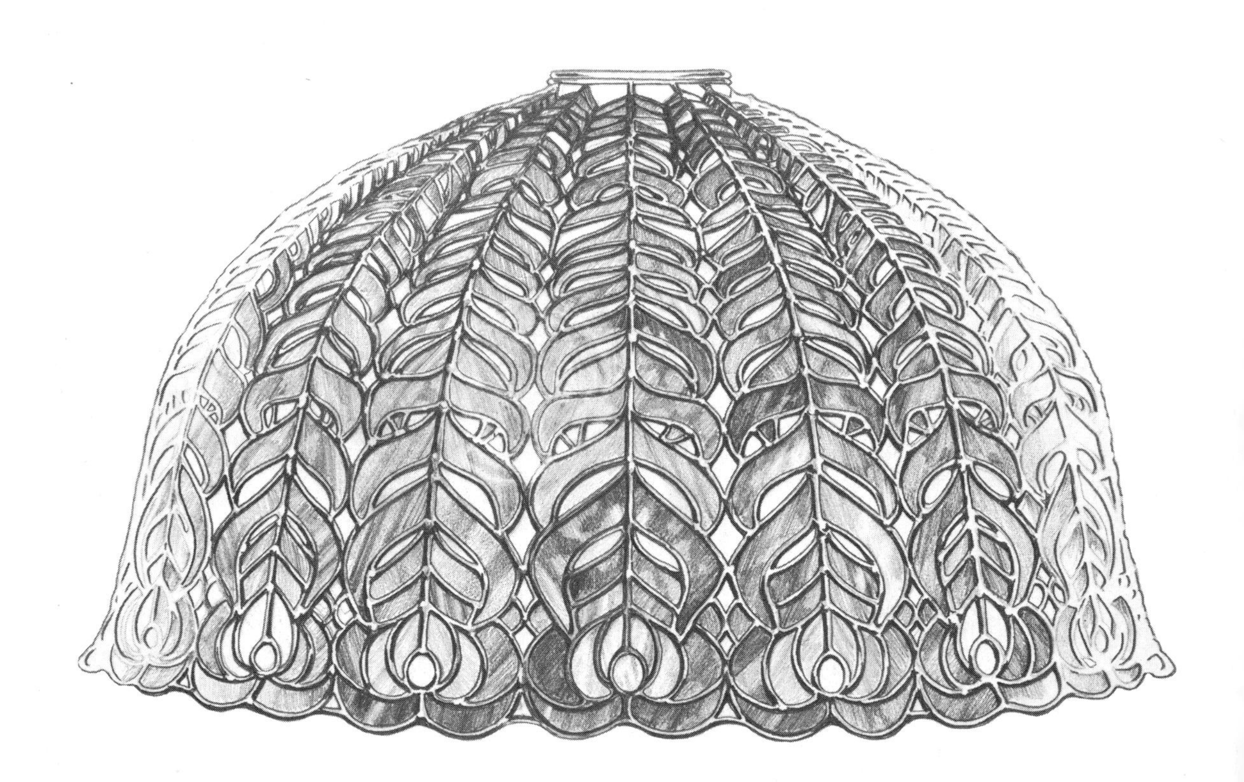

LEADED GLASS

a handbook of techniques

by
Alastair Duncan

Watson-Guptill Publications
New York

First published in the United States in 1975 by Watson-Guptill Publications, a division of Billboard Publications, Inc., 1515 Broadway, New York, New York 10036

First published 1975 in Great Britain

Manufactured in Great Britain

Library of Congress Cataloging in Publication Data
Duncan, Alastair, 1942—Leaded glass.
1. Glass craft. I. Title.
TT298.D86 748.5 75–1037
ISBN 0–8230–2660–4

There be none of Beauty's daughters
With a magic like thee.
Byron

Tiffany pointsettia lampshade
The Egon and Hildegard Neustadt Museum for Tiffany Art, New York

Contents

Note As the glass industry is still working to imperial measurements these have been maintained through the book without conversions being given.

Acknowledgments

My thanks for help and advice in the compilation of this book are as geographically widespread as they are numerous.

In England, to Caroline Swash, Honorary Secretary of the British Society of Master Glass-Painters; C. J. Thwaites, Chief Metallurgist at the Tin Research Institute; and glass designers Alfred Fisher and John Lawson for their comments on various sections of the text. To Christopher Salmond of James Hetley and Co, Wembley, for his co-operation in making available to the public the materials described in the book; to the staff of Goddard & Gibbs for their advice; to Jim Farnill for his photography and Robin Lawrie for his illustrations; and, especially, to Thelma Nye and William Waller who, in editing the book, provided a continual source of encouragement and direction.

In America, on the west coast, to Peter and Joanne Nervo in Berkeley, California, for all their help and friendship and to whom this book is dedicated as a small token of my gratitude; and to Dan Fenton, Editor of Glass Art magazine, in Oakland, for his supply of photographs. On the east coast, in New York, to Dr. Egon Neustadt, for showing me over the Tiffany museum; in Brooklyn, to Robert Sowers, for his advice; in New Jersey, to Dr. Seymour and Anita Isenberg, for their pioneership in the revival of the art of glass and for their book *How to work in Stained Glass*; and in Washington, D.C., to Walter Hamilton and Sal Fiorito for their kindness.

Thanks, also, to all the glass designers, artists, and collectors in America, England, and Germany, who have contributed photographs of their work and collections for incorporation into the book. The photographs of Ed Gilly's work were taken by Darleen Rubin.

My gratitude, finally, to Betsy Joffe, who did so much to get the book underway.

Foreword

There is often a certain amount of confusion in people's minds as to the difference between the concepts of 'art' and 'craft', and 'artist' and 'craftsman'. My own thoughts on forming a distinction between the two are that all art involves craft, but that not all craft necessarily involves art.

As far as this relates to the subject matter of this book, leaded glass, I feel that the choice of whether one is a glass artist or a glass craftsman depends on the amount of originality in the finished object. A glass artist is someone who creates his *own* design and colour balance, and then executes them in glass; a glass craftsman, on the other hand, is someone who takes an *existing* idea and then translates it into glass.

This is an important distinction as it underlies the fact that it is not the medium *per se* – as is so often incorrectly assumed – that determines whether something is an art or a craft, but rather the amount of creativity applied to that particular medium. This puts the onus squarely on the glass exponent – and not on the glass – as to whether he or she is an artist or craftsman.

The following pages cover the technical, ie craft, aspects of working with glass, but they are intended, also, to instil in the reader a sense of its phenomenal artistic potential.

London 1975 A D

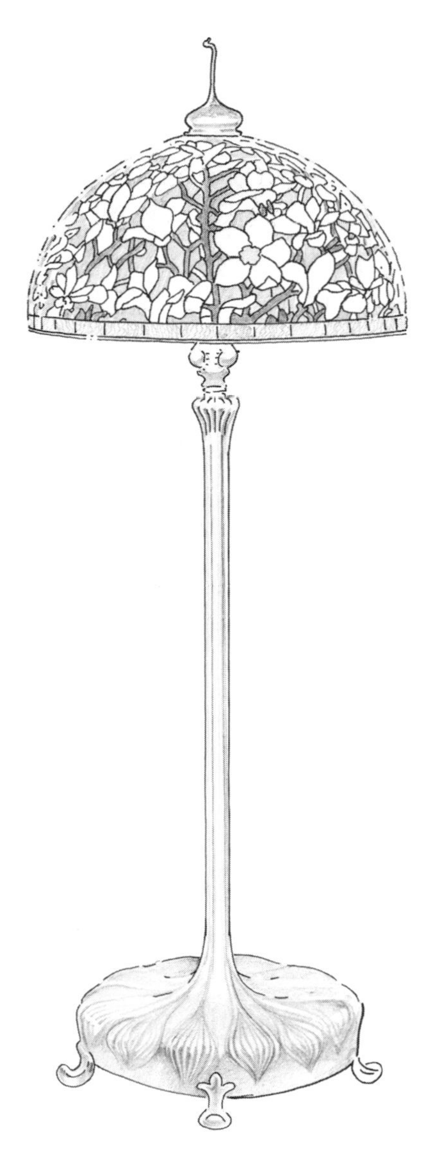

Introduction

My reasons for writing this book are two-fold: first, to introduce the art of glass craft to the uninitiated, and second, to offer, I hope, new interest and stimulus to the already established glass craftsman.

Glass, like any other art form, needs a constant influx of new ideas, talents and energies, and I hope to fire the beginner with enthusiasm for working with glass, as, although it is one of the most versatile and rewarding of all the art media, being both decorative *and* functional, it has become somewhat of a 'lost art'.

It is important, initially, to distinguish between leaded glass and stained glass. The term 'stained glass' can be confusing and is often a misnomer when related to working in glass. It tends to refer to all work done with coloured glass, whether stained or not. The process of staining glass, ie applying a special stain to the glass and then firing it, falls outside the scope of this book. There is, though, a great deal of literature on the technique of staining glass; a selection is given in the Bibliography.

For the purposes of this book I will deal purely with leaded glass, which involves the use of glass that has received its colouring during manufacture and which will not be further stained, painted, fired, etched, or otherwise treated. The glass is, therefore, purchased in its finished form, ready for immediate use. None of the projects shown in the following pages entailed any processing of the glass after it has been selected and bought (other than, of course, to cut it into its required shape for the design at hand).

It is this that makes leaded glass ideal for the home craftsman. There is no need to acquire the technical and artistic skills necessary to paint or fire glass, or to buy any expensive and space-consuming equipment such as kilns. All that is required is a small work area in which to cut and assemble the glass.

At any one time only two of a total of three materials will be needed to make up projects; either glass and lead came, or glass and copper foil. This is phenomenal if one thinks of the diversity of shapes, designs, and polychromatic effects that can be obtained with just two materials. A modern alchemist, working likewise with only two base materials, could not transmute them into anything more beautiful. It is this confinement of leaded glass to only glass and lead came or copper foil that makes the craft so readily accessible to the home craftsman.

There is frequently, for the layman, an aura of impenetrability surrounding most crafts. One often feels that they are beyond one's capabilities. A typical reaction, on seeing a work of art, is one of disbelief that one could do it oneself. However, glass is not the sole preserve of the commercial stained-glass studio. Top-quality glass work can also be produced in the confines of the home. Obviously the beginner has to familiarize himself with the various tools, materials and techniques of the craft, and whereas it may take some time to reach a preferred level of proficiency, this is a prerequisite of any worthwhile pursuit. There is, though, no fundamentally difficult aspect of working with leaded glass. Just as the amateur has transcended this fear of the unknown in other crafts – such as pottery, leatherwork and jewellery – so he can easily do it in glass. In America the home craftsman's interest in glass has, in recent years, reached boom proportions, generating feverish home industry and some monumentally fine work. Some examples of such, from a number of the finest glass exponents in both America and Europe, are shown in the following pages, a large percentage of which was done, not in a professional studio, but in the craftsman's home.

By far the largest percentage of professional work in glass in England is in the traditional field of religious glass, whether in churches, chapels, or synagogues. Due, however, to the marked decrease in the number of churches being built, and the increasing number being demolished (existing churches supply a not inconsequential amount of replacement and restorative work), the number of professional stained glass studios has, due to this fall-off in commissions, shrunk accordingly. In addition, most churches have limited funds with which to commission windows, and this has been a further constraint on the potential growth of the stained glass profession. Unless there is a rapid expansion of glass beyond the ecclesiastical realm into the more secular ones of commerce, industry, and the home, stained glass work will languish and gradually disappear.

This is in no way to suggest that there is not a vital and legitimate role for church-commissioned work in the range of forms that glass can take as an artistic medium, but the reliance, or, rather, overemphasis, on this one outlet until now has stifled its growth in other directions. What is required, therefore, is a shift in balance: an expansion into other areas both to compensate for the fall-off in the traditional market for stained glass, and to expand more into what is, in any event, part of its rightful preserve, ie, the office, the home, the factory and the school. It is a propitious moment for the English glass practitioner to follow his American and European counterparts in this respect.

One of the most effective ways of extending the utility and creativity of glass is to add another dimension to its traditional use. Glass is, because of its innate properties of luminosity, potentially the most innovative and aesthetic of all the art media, yet it has not yet been fully utilized in England, due, largely, to the glass craftsman's almost exclusive concentration on two-dimensional glass such as windows and panels, at the almost total neglect of glass as a three-dimensional form of art. The extra dimension helps to extend the effectiveness of glass beyond the hours of daylight which, if it were to be the only form of illumination available, would be a limiting factor on the more widespread use of glass as a functional art form. Whereas natural lighting (the traditional means of illuminating two-dimensional glass) gives a beautiful effect due to its continual variation in intensity throughout the day – thereby producing an ever-changing diffusive and kaleidoscopic effect on the coloured glass through which it is transmitted – its magical effect is suspended between dusk and dawn. Three-dimensional glass objects, such as lamps and lanterns, do not, however, have this reliance on solar energy. They are backlighted by artificial lighting which extends the function and beauty of glass right around the clock. It is in this respect that I hope this book will be of use to the established glass craftsman: to add, quite literally, another dimension to the existing range of glass that he produces. Two materials, which have not until recently been manufactured in England, are now commercially available. These are U-frame lead cames and copper foil; the non-availability of which has, until now, largely precluded any serious attempt to produce intricate three-dimensional glass objects such as lampshades.

Glass as a non-religious art medium is, in fact, at present almost totally non-existent. So little work has been done in glass for the general public for so long that they are no longer aware that it is a major art form. It is difficult to know which is more alarming – the fact that there is, for example, a ready market for plastic-moulded imitation glass lampshades, or that most people who buy them are unaware that they are, in fact, copies of the art of another, more creative age.

This book is an attempt to remedy the situation by introducing the technique of leaded glass to a wider audience than at present exists.

Chapter 1

Traditional stained glass, Tiffany and the present revival

Let the Past serve the Present Mao Tse-Tung

This chapter gives a brief history of leaded glass as it relates to the materials and methods employed in this book. The subject does, of course, warrant a complete book on its own, but I have tried to limit it to the major events that have influenced the course of its history so that it can act as a frame of reference for those readers new to the craft. It should be stressed that a knowledge of the history of leaded glass is not prerequisite for any of the technical aspects that are covered in this book. It is not unreasonable, however, to assume that an awareness of the craft's past will lead to a sense of familiarity with and confidence in what one is doing – as to where one 'fits' in the overall picture – and it is for this reason that I have incorporated it in this book. The history is divided, for simplicity, into two main sections: the traditional, two-dimensional method of utilizing leaded glass in window or panel form; and the more recent method of using it in three-dimensional shapes.

Two-dimensional glass

The origin of leaded glass is shrouded in obscurity, but its beginnings were in all likelihood Egyptian, some time between 7000 and 2000 BC. Whereas it was possibly first used as jewellery, it was later incorporated into window-form to fit into the small holes that were made to let light into buildings. However, as the techniques of glass-making then only allowed for very small pieces to be cast in a single operation, so the architects probably had to combine several pieces to make up the dimensions of the hole. This, then, was the origin of the geometrically patterned windows that are assembled by today's glaziers: a modest forerunner to the magnificence of Chartes, perhaps, but one whose technique is, as yet, fundamentally unchanged.

From Egypt the use of glass spread gradually via the levant to Byzantium and then westwards to Greece and Rome, and it is known from excavations at Herculaneum and Pompeii that the Romans developed, or used existing, methods of filling circular openings in their stonework with pieces of glass, and also of using lead or copper frames to hold the pieces in position.

The craft developed steadily during the next millennium: glass-bowing replaced moulding; im-

An early example of flat glass
Glass plaque inscribed with the name and affiliation of Ptolemy; hieroglyphics on one side, Greek on the other. 220 BC
Department of Antiquities, British Museum, London

proved technical skills led to the introduction of transparent glass in place of the previously more opaque, mosaic variety; and the technique of painting on and firing glass likewise progressed. From this point on, the history of leaded glass in two-dimensional form is largely that of the history of stained glass, as virtually all leaded glass used in this form is painted.

Stained glass was used throughout this period almost exclusively as an ecclesiastical art form. Not only did the glass act as a source of light for the dark interiors of churches – and keep out the elements – but also the windows were designed to form a pictorial representation of a more celestial world: to take the congregation's mind off the hardships of their own lives and to contemplate the divine. Stained glass acted, therefore, as a form of iconography which depicted Christ and Christendom's leading personages by means of images painted on brightly coloured glass.

Detail of an early Gothic stained glass window; twelfth century
Canterbury Cathedral, England

The craft really came of age in the eleventh century and progressed rapidly throughout the Gothic period to reach its high point in the thirteenth and fourteenth centuries. Perhaps the oldest stained glass windows are in the clerestory of Augsburg Cathedral in Germany, which are dated as 1065, while Le Mans, Chartres, Canterbury, York and Gloucester followed later.

There are some exquisite examples in the above mentioned cathedrals of how best to juxtapose pieces of differently coloured glass artistically; how best to incorporate leading in a design; and how to paint on glass (as distinct from the technique of painting on canvas; a difference which even though this book does not cover any aspects of painting on glass, is still of relevance as it underlines the fundamental design function of the lead as it holds the pieces of glass together). No sooner had the glass practitioner mastered his art than political forces were brought to bear (particularly in England) which largely relegated stained glass to the backwaters of creative art for some three hundred years. Luther's Diet of Worms (1521) challenged papal

infallibility and led to the Reformation which, in England, caused the Protestant Tudors to likewise rebel against Rome. Stained glass windows became part of all that was a manifestation of Catholic omnipotence, and which led, by Royal decree, to a spell of iconoclasm in which a great deal of the finest glass work was systematically destroyed. Henry VIII's injunctions of 1547 read, in part:

> Also, that they take away, utterly extinct and destroy all shrines, all tables, all candlesticks, trindles and rolls of wax, pictures, paintings and all other monuments of feigned miracles, pilgrimages, idolatry and superstition so that there remain no memory of the same in walls, glass-windows or elsewhere within their churches or houses.

Such Governmental Ordinances acted as a deterrent to church-commissioned windows, and the craft orientated itself towards heraldic and domestic windows to offset the fall-off in religious work.

Then, as if this was not in itself bad enough for the stained glass craft, the mid-1600s saw the rise of Cromwell and the Puritans (1646). They, in turn, saw brightly coloured glass as gaudy and self-indulgent, and therefore heretical. Further 'purification' followed.

Political forces not only proved disastrous for existing stained glass, but also for its manufacture. On the continent the technique of making coloured glass had become a highly specialized and centralized industry. Virtually all glasshouses were situated in Lorraine as the forests bordering the Rhine provided an ample supply of the principal raw materials required – fuel for the furnaces and wood ashes containing the alkali necessary for making the

Contrasting styles in stained glass. On the left, Sir Joshua Reynold's 'Fortitude' window in New College Chapel, Oxford. On the right, a section of the window to St Cecilia in Christ Church, Oxford, by William Morris's Pre-Raphaelite colleague, Edward Burne-Jones. Reynolds used enamel paint on clear glass and employed the leading in a totally functional grid format. Burne-Jones painted on coloured glass and incorporated the leading so that it formed the linear detail of each separate aspect of his design

glass – with the river itself providing transport for the glass to neighbouring countries, especially England. In 1633 Lorraine was devastated by war. The glassmakers fled, and their furnaces were razed to the ground. This brought the major source of coloured glass to an end, with the result that stained glass artists tried to improvise by painting white glass with colour vitreous enamels. The mediaeval technique of using glass as a pictorial art form which had made full allowance of the art's chief characteristic, transmitted light, was, therefore, lost. Whereas the use of enamel paints could provide the desired colour effect, this meant that – if the colour could be obtained by means of the pigments in the enamel – so there was no need for coloured glass, which, in its turn, meant that the end-result was no longer translucent. Church windows became, in effect, large opaque paintings. New generations of glass painters came to think of stained glass in terms of a painting designed independently of the glass, and merely applied to it.

By the eighteenth century the craft had reached its nadir with artists moving from canvas to glass with no adjustment being made for the change in media. A classic example of this is the west window of New College Chapel at Oxford, which was designed by the celebrated portrait painter, Sir Joshua Reynolds, in 1778. The end-result would have been magnificent as an oil painting, but, copied by the glass painter in coloured enamels on squares of ordinary window glass, I feel that it is totally unsuited to a medium that is meant to utilize transmitted light rays to enhance its overall effect. In addition, the lead came's function is wholly unartistic: it provides solely structural support for the glass. It would have been much simpler and more effective to have walled up the window and painted a mural.

Not that Reynolds alone deserves such censure: many other less renowned artists were guilty, if not of using enamels, then of employing a chiaroscuro light and shade technique – once again covering the glass with paint and sealing off the light.

The rediscovery of the lost formula for making mediaeval coloured glass, now commonly known as 'antique' glass, was undertaken by two Englishmen: first by Charles Winston in 1850, and then later by William Chance. They realized that the success of glass painting depended as much on the quality of the material employed as on the skill of the artist, and that it was necessary, therefore, to obtain glass similar in colour, texture, and tone, to that contained in Gothic windows. Several years of experimentation and research were crowned by eventual success: the antique glass produced being, in fact, superior in quality to the twelfth and thirteenth century variety that they had set out to imitate.

It was left, finally, to William Morris in the 1860s to revive the mediaeval glass craftsman's expertise by incorporating the new glass in his windows. It is indeed fortunate for the craft of stained glass that a man of such protean talents should have included it in his range of art forms, as his prestige gave the craft a much needed renewed legitimacy. Not only did he and his fellow Pre-Raphaelites, Rossetti and Burne-Jones, design stained glass for ecclesiastical purposes, but also they popularized it by extending its use into the domestic sphere.

After the brief innovative spirit shown by Morris (whom I shall be returning to shortly as he was the precursor of the art movement that led to the use of glass in three-dimensional form) the craft returned once more to its tradition-bound role; with most work being commissioned for religious purposes, although coloured glass did undergo a brief popularity in heraldry and Victorian door panels and fan-lights. This criticism, incidentally, is aimed mainly at English glass, as its history on the Continent has shown, especially in this century, a far more adventurous spirit, with the last 25 years in Germany, in particular, producing many superb windows by designers such as Schaffrath, Schreiter, and Meistermann.

This brief summary of the events that have highlighted the craft's past, is, as such, a generalization. There have always been, even when the average level of creativity was at its lowest, some very fine glass exponents whose work remained untainted by the general prevailing standards. A great deal of this has, though, not survived the ravages of time, politicians, and vandalism.

The history has, therefore, been one of vicissitudes – with rather fewer ups, perhaps, than downs – but the present situation is one in which I feel it is on the threshold of a very marked upswing. The reason for such optimism is two-fold: first, that the Individual (as against the Firm) is at last finding, mainly in America, that the demand for both painted and unpainted leaded glass (not just church window glass, but commercial and domestic glass as well), has increased to the stage where it is financially self-supporting, without the individual having to rely in addition on other forms of income. This has a self-perpetuating effect, as the more glass work that is commissioned, the more the public is exposed to it, and the more then that they subsequently demand. Second, recent legislation in Germany has auspicious connotations for leaded glass. A law has been passed which requires that a certain percentage of all money invested in new architecture must be spent on art work, and it is likely that similar legislation will follow in other countries.

Both of these points are good omens for leaded glass craft in general, as the sooner it becomes firmly established in one country, the sooner it will spread to others.

German glass designers have taken the lead in unpainted ecclesiastical windows. A church window of modern design: outside view of the east wall of the St Marien Catholic Church in Bad Zwischenahn, Germany, illuminated from the inside
Ludwig Schaffrath, Alsdorf-Ofden, Germany

Three-dimensional glass

William Morris was influenced by Ruskin's attack on the prevailing mid-nineteenth century traditional, or, as he termed it, Neo-Renaissance, style of art, and with his Pre-Raphaelite Brotherhood friends started a counter-movement back to nature. (A similar pantheistic movement occurred simultaneously in all the arts, being directed mainly at the new Machine Age and the stultifying effect that it had on music, literature, painting, and poetry.) Morris studied nature and extracted its common denominators – its curvilinear motifs and botanical shapes – which he used as the central theme in his designs, and which, in turn, became the *leit motiv* of the movement which he was to precipitate; namely, Art Nouveau.

Art Nouveau, which began in England but was to reach its zenith in France, was an ornamental style in which the intention was to capture the quintessence of nature in its every form. Everything had to depict the undulating and biomorphic shapes and lines of flowers and plants. The movement started in the 1870s and reached its full momentum at the turn of the century (often, in fact, termed as *fin de siècle* art), generating some of the finest art the world has ever seen, and leading, as far as we are concerned, to a resurgence in the use of leaded glass as a decorative art form, but with one important difference: that it was for the first time effectively incorporated into three-dimensional form. That this happened is due in part to, first, Swan's (1878), and then Edison's (1879), inventions of the incandescent filament bulb, which they showed independently at the great Paris Universal Exhibition in 1889. Up to this stage lighting was provided primarily by liquid fuels and by gas, both of which had limiting design constraints. The functional requirements for liquid fuels, such as methylated spirits and paraffin, were that lamps had to be easily accessible so that they could be refilled regularly, that they always be placed in a strictly upright position, and that they have a protective cover and a fuel reservoir.

Gas, on the other hand, although giving a good incandescent light, had to be in a fixed position and attached to a pipe which, although it could be somewhat disguised, spoiled the appearance of chandeliers and wall-lamps. The above were factors that had obviously limited the design aspects of lamps up to that time.

With the advent of the electric light, the safety considerations which had affected all forms of combustion lighting disappeared, and lamps no longer had to be vertical. This allowed the Art Nouveau modernists to turn their creative talents to the new design possibilities of lighting.

Gallé, Lalique, and the Daum brothers in France, Tiffany in the United States, and several other 'leading lights' designed and created glass lampshades which, while faithfully adhering to the stylistic traits of Art Nouveau, made full allowance of the new-found freedom of innovation that the electric light afforded.

Door panels in typical Art Nouveau style showing the characteristic 'whiplash' curve. Unsigned, probably early 1900s

Reproduced by courtesy of Bruce Rabbino, Manhattan, New York

All these artists worked with blown glass (examples of which are on permanent display at the Musée des Arts Decoratifs at the Louvre and are well worth seeing), but it was Tiffany who diversified to develop a method of producing machine-made glass in flat form. His incorporation, in addition, of copper foil to bond pieces of glass together, which he used because the traditional method of using strips of lead (called 'cames') was too unwieldy and insufficiently pliant for three-dimensional glass work, allowed him to use his flat glass to make curved lampshades. It was Tiffany, therefore, who was the founding father of leaded glass in three-dimensional form.

1848, the year that spawned so much revolution throughout the world, was to see the birth also of someone who was to revolutionize the world of glass: Louis C Tiffany. Tiffany was the son of the most fashionable American jeweller, and whose shop, Tiffany & Company, still exists on Fifth Avenue, New York (it opened a branch in Regent Street, London, in 1868). He spurned his father's offer of a place in the firm, however, opting rather for an independent life in the decorative arts, and he was one of the first to be drawn to Europe by the reputation of William Morris.

Tiffany showed a remarkable versatility throughout his artistic career, using numerous different materials to produce a wide range of *objets d'art*, but he is remembered primarily for his lampshades and, more specificially, for the unique glass with which he made them.

Tiffany's search for a new kind of glass had stemmed from his dissatisfaction with the coloured glass in America at that time: the largely monochromatic, transparent variety used in stained glass windows. Multi-coloured effects could be obtained only by the juxtaposition of the pieces and, in many instances, by the addition of paints and stains. For three main reasons he found that this was inadequate for his needs. Firstly, for an adherent of the Art Nouveau style, that the monochromatic character of the glass was untrue to life, since nothing in nature is uniform in colour; secondly, that the application of paint interfered with the transmission of light, one of the principal functions of glass; and, thirdly, that the glass lacked the innate properties of iridescence and radiance that he desired.

After innumerable experiments he finally obtained the correct chemical interaction and, in 1881, acquired the patent on an iridescent glass. This was followed by more experimenting to perfect the method of creating multi-coloured, opalescent glass, and his efforts were rewarded in 1894 when a further patent was granted for his *Favrile* glass. The word *Favrile* derives from the old English *fabrile* meaning 'handmade', which he changed to *Favrile* to create a unique word for the trademark of his Tiffany Glass and Decorating Company, which he had established in Corona, New York.

That the name *Tiffany* is now generally used to describe all leaded glass lampshades is a tribute to his monolithic influence on this art form, but it is, in effect, a backhanded compliment as nobody else has ever managed to match his creative brilliance

French lamps were not made with leaded glass: A selection of floral designed glass lampshades and vases by Emile Gallé, 1846–1906, which consist of layers of coloured glass laminated together in cameo fashion. To obtain the desired effect, the design on the glass was covered with wax, and the remaining areas were acid- or wheel-etched to different depths, thus giving the glass a relief surface and various shadings of colour
Reproduced by courtesy of Galerie Maria de Beyrie, Les Halles, Paris, France

in glass- and lamp-making. Anyone who has had the opportunity to see an original Tiffany will appreciate just what it is that made his work unique. On the one hand, his glass was impregnated with a special combination of metallic pigments often in striation or mottle form which, although sometimes fairly flat in colour when viewed in reflected light, would come alive in transmitted light; being variously kaleidoscopic, incandescent, or phosphorescent. Added to which, the glass with which he worked was often textured in nodular, rippled, or fibrillated form to create extraordinary light refractive effects.

In addition, his designs were singularly his own. Not only did he incorporate the prevailing Art Nouveau style into his work to produce floral lamps such as the tulip, apple blossom, jonquil, and pointsettia designs, but he also made a variety of geometrically shaped shades. Between the two – his glass and designs – his work has remained unparalleled to this day.

Art Nouveau, like Baroque and Rococo before it, was in its turn to be replaced, this time by such art movements as Surrealism, Art Deco, and the Bauhaus, and it has only recently undergone a renewed popularity. Tiffany, whose work till five years ago had seemed destined largely to the world of the museum and the antiquarian, is suddenly back in fashion, and prices of his lamps have consequently rocketed. A Tiffany Wisteria lamp which, for example, sold originally for $400 in 1906, was auctioned at Sotheby Parke Bernet in New York in 1974 for $42,000.

The revived public interest in Tiffany in America has contributed, in part, to the renaissance that the craft of leaded glass is experiencing there at this moment. Not only are Art Nouveau styled lamps being made by professionals and amateurs alike, but the craft has diversified to incorporate innumerable other forms of both two- and three-dimensional glass: free-form objects, sculpture, pot-plant holders, mirrors, room dividers, etc. The craft has, in fact, exploded into a boom industry, and we are on the threshold of a new, and very distinct, chapter in the history of glass, one which is drawing on both traditional and contemporary art styles to make the craft really eclectic for the first time ever. An exciting range of old and new floral and geometric designs are being reproduced in glass, and as far as style is concerned, anything goes, and often with bewildering effectiveness.

Left
Tiffany's famous Wisteria lampshade, of which few remain. [From 'The Lamps of Tiffany', by *Dr E Neustadt*] *Courtesy of the author*

Right
Unsigned floral lamp by Handel, a contemporary of Tiffany known more though, for his painted, rather than unpainted, shades. The Handel Corporation of Meridea, Connecticut, was founded in 1885 and was operational till 1936

Chapter 2

Glass

The first material with which it is necessary to become conversant in leaded glass is, of course, the glass itself. Knowledge of the different kinds of glass commercially available – not only in sheet form, but as jewels, bullions, roundels, etc – and which kinds are best suited to which particular form of leaded glass work, will enable the craftsman to maximize its potential as an art form.

Flat glass is now made not only in North America, England, and Europe (its more traditional spheres of manufacture) but in such globally diverse countries as South Vietnam, Mexico, and Japan, and the accumulated range of colours, textures, and degrees of translucency now produced are delightfully confusing. Not, of course, that any one supplier can carry a complete, or nearly complete, selection of all available domestic and imported glass, but most suppliers – those, that is, who cater for the glass artist and his needs for decorative glass – offer a sufficiently wide range to meet most needs.

It is not necessary for a leaded glass exponent to concern himself unduly with how glass is made as he buys it in its finished form, ready to be cut and assembled. It is, nonetheless, an interesting process which warrants a brief mention as the more the craftsman knows about the glass (and other materials) that he is using, the more he will be able to recognize and appreciate differences, however seemingly insignificant, in the properties of similar sheets of glass and the more acuity he will develop in his subsequent choice for the project in hand. Connoisseurship in glass selection can only come with an intimate knowledge of what is available.

Early glass blowing
Le Viel 1747

Handmade glass – as distinct from the machine-rolled variety; a different process – consists of a base compound of, for example, sand, soda potash, lime, and barium (the ingredients and method of glass-making often vary between manufacturers), which are riddled, first separately, and then together, in a fine sieve. Certain chemicals, depending on the required colour of the glass being made and also, partly, on its translucency, are added to the base compound, and the whole mixture is heated till it melts into a glassy mass at about 1400°C. Then the mixture is gathered from the pot with a blowing iron, and blown into its penultimate shape (usually cylindrical), prior to being cut open. It then undergoes a lengthy annealling (cooling) process during which the glass is finally flattened out into sheet form.

Two different surface textures; on the left an American ripple finish and on the right a Flemish wavy one

Glass's trinity: its colour, translucency, and texture

Colour

It is the chemicals, usually metallic oxides and sulphides, which act as the pigments which, in turn, form the range of colours on which we are so dependent for artistic effects. For example, manganese oxide produces a purple coloured glass; cobalt a blue; cadmium sulphide a canary yellow; and selenium a red. Chemicals can, in addition, be impregnated into the glass not only singly, but in combinations to form either a new monochromatic hue or alternatively polychromatic patterns.

These latter, multi-coloured patterns within the glass itself, are usually manufactured in striation form. In a striated sheet, known as a 'Streaky', the colours form bands or lines that flow at random across the sheet to form waves or whorls, giving the appearance of long, rhythmic brush strokes, and often producing a close approximation to nature itself. More will be said about striations in chapter 7 on *Design*, but remember at this stage that a 'Streaky' glass is innately interesting due to its variegated colour effect and that it is, as such, one of the kinds of glass that will probably be used a great deal in compositions. As a leaded glass exponent you will not paint, stain, etch or otherwise treat the glass after you purchase it, and you must, therefore, make full use of the glass's intrinsic qualities to help obtain your desired artistic effect: try, therefore, at all times to be conscious of making the glass work for you within your design. It has to provide all the spectral effects that you wish to achieve.

Translucency

A second aspect of glass of which to be aware is its degree of translucency. Coloured decorative glass ranges in density of colour from almost transparent to semi-opaque (when the term *opaque* or *opalescent* is used in leaded glass work it is understood to describe glass in which the transmission of light is considerably reduced but not totally eliminated: a glass that is completely impervious to light is of no aesthetic use to us). Whether to choose glass that transmits a little or a great deal of light depends on personal preference, which should, in turn, be influenced largely by the end purpose of the project in hand. There are no hard and fast rules in this, though suggestions as to what colour density of glass is best suited to what particular form of leaded glass work are given in chapters 7 and 10; the decision on how best to look through a glass

(overleaf)
Detail from the ceiling at the Maxwell Plum restaurant in Manhattan. It comprises 126 panels of glass each two square feet.
Courtesy Maxwell Plum Restaurant.

(right)
Passion Flower
Kathie Bunnell, Mill Valley, California.

(below left)
A section of Ed Gilly's famous bed canopy which consists of hundreds of circles and ellipses of transparent coloured glass held together by copper foil in an undulating mushroom shape.

(below right)
Unpainted Tiffany landscape window.
Courtesy Dr. Egon Neustadt, Manhattan, New York.

depends, in general, on whether the project being made, for example a window or lampshade, is to be more functional than decorative, or vice versa.

Texture
The third characteristic of glass that can be utilized creatively is its texture, both on its surface and internally. A great deal of the coloured glass that a leaded glass practitioner uses is fairly uniform in thickness, with its top and bottom surfaces smooth, ie with no textural qualities at all. But some do have textural qualities. It is important to be aware of these and to know when best they can be incorporated into a design so as to add sophistication and sparkle.
Surface textures Some machine-rolled American glass is patterned (ie on one side only: one side of a sheet of glass must always, of course, be made relatively smooth during manufacture so that it can be scored with a glass cutter) in various configurations, such as a granular, nodular, or rippled form. Each one of these not only gives the glass a tactile quality, but modifies the transmission of light in its own unique fashion so that the light rays, on striking the uneven surface of the glass, are refracted through it at different angles. The result can greatly enhance the overall effect of the glass, giving it a dazzling aliveness.

English antique glass also contains various surface textures. It has, for a start, a wavy surface (due to the fact that it is handmade and not machine-rolled), which in itself often gives a reamy finish to diffuse the light. Alternatively, a deeply incised and irregular pattern, termed, aptly, 'Curious' antique, is applied to the surface of the glass to produce a different, and very marked, diffusive effect.

Internal textural patterns are less frequently made nowadays. (Tiffany used to make a spectacular 'fractured' glass consisting of splinters or flakes of differently coloured glass superimposed on top of each other and pressed into a single sheet, which, unfortunately, none of today's manufacturers has tried to emulate.) The only one that I know of is the 'seedy' antique. This contains tiny air bubbles inside the glass itself which have a far less pronounced effect on the passage of light through the glass than most surface patterns, and so are best incorporated into a design when the desired visual effect is one of subduedness or subtility.

Varieties of glass

The following are some of the more popular varieties of decorative glass usually stocked by those glass suppliers who serve the glass artist. Every stockist carries a different range of glass, consisting usually of a complete range of the domestic varieties and a selection of some imported ones. The first thing is to go to your supplier and familiarize yourself both with what he does stock and what he can, in addition, order for you. Only a limited amount can be learnt about the different properties of glass by reading: you really must both see and see *through* them for yourself to determine how best they will meet your creative needs.

1 *Antique glass*
This is a handmade glass manufactured in a wide range of forms, the most common of which are:
(a) *Pot glass* This is glass of a single colour which is known as 'pot' glass because it only needs to be made from one pot of one colour (a pot is the vat in which the molten glass is kept in flux while the antique glass is being blown).

The beauty of pot glass is in its tonality and degree of saturation of colour. Because it is monochromatic it is perhaps less interesting on its own than multi-coloured or textured glass, but it can, when juxtaposed selectively – such as to provide stabs of brightness amongst more neutral tones – provide great richness of colour.
(b) *Streaky antique* This consists, as its name implies, of a basic colour with streaks of another or other colours running through it. The streaks are irregularly patterned due to their having been created by mixing two or more pots of glass

Detailed photograph of the internal properties of an English 'seedy' antique glass

together during manufacture, and the resulting effect can be vividly kaleidoscopic.

English streaky glass is transparent and is, therefore, ideally suited to windows and other objects that need to transmit a large amount of light. Try, when using it, to keep the pieces of glass within the design as large as possible: pieces that are too small will lose the continuity of the band effect of the colours as they swirl across the glass.

Glass panel clearly showing the striations in English Antique 'Streaky' glass
Ernest Porcelli, Brooklyn, New York

A sheet of opalescent glass showing the bands of colour that it contains

(c) *Flashed glass* In this variant of antique glass a thin layer of coloured glass is fused with another, thicker base colour, usually white or clear. The result is, in effect, two sheets of glass laminated together such as, for example, flashed ruby on white.

Flashed glass is of less value to us than to a stained glass artist, who can use the double layer of glass to his advantage by etching the coloured layer to obtain various effects, but it is made in several colours not offered in pot or streaky glass and does, therefore, extend the range of glass available.

2 *Opalescent glass*

This has always traditionally been manufactured in the United States, but is now also made in such countries as Germany, Belgium, and, more recently, in South Vietnam.

Opalescent glass is made in a number of ways: as a single colour: polychromatically, with the pigments mixed to give the glass a streaky, mottled, or cloudy appearance; and with or without a surface texture. It can be both a most beautiful and challenging glass with which to work, and the demand for it does, because of this, often outstrip the available supply, so pay close attention to glass economy when using it. In this respect it is a good idea, when selecting the multi-coloured variety from your supplier, to purchase at least 50% more than you anticipate that you will require for the project at hand. This is because the pigments are mixed into opalescent glass by hand during manufacture with the result that the colour patterns and tones in the glass are never exactly the same in any two sheets, and if you experience difficulty in cutting the glass, thereby incurring unexpected wastage, you may

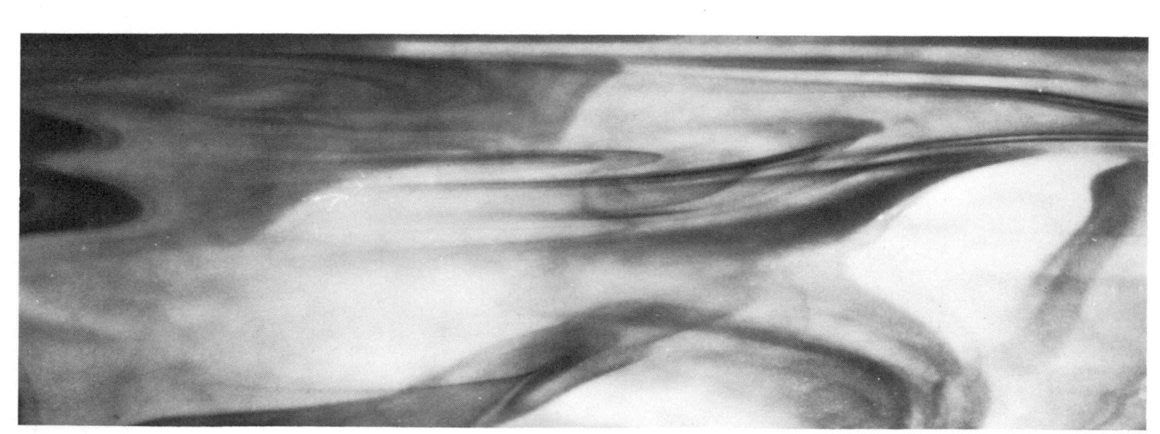

then find that you have run out of the particular colour pattern that you are using. And you will invariably find, maddeningly, on returning to your stockist, that you cannot match your original purchase because somebody else has, in the meantime, bought the section of the sheet or sheets that you left.

Opalescent glass has, I feel, one characteristic that transparent glass does not: namely, that it can be seen by both transmitted *and* reflected light. If you look at, say, a church window from the outside or at night when there is no transmission of light through the glass, then it appears flat and colourless and, therefore, ineffectual. This is due to the fact that transparent glass needs to be illuminated from behind to highlight its innate qualities. Opalescent glass, however, has its colour impregnated into it to the extent that the pigmentation is visible by light rays reflecting off it. It can be seen, as well as seen through. This means that opalescent glass is potentially decorative for 24 hours a day (against this it should be noted, however, that the more light that a particular glass object, such as a window, is required to transmit, then the more opacity in the glass the less functional it will be. Always select your glass to suit its end-purpose).

3 *Commercial glass*
Clear glass should not be eliminated, ie that used normally in office and home windows, from your glass repertoire: it is sometimes the very absence of colour and texture – the banality of clear glass – that enables it to be used extremely well on its own or in conjunction with antique or opalescent glass in windows. Too much streaky glass, for example, can be visually overpowering, so in order to prevent such diminishing returns, the composition can be balanced with clear glass, which often fills in ideally as a background or neutralizer.

4 *Mirror glass*
It is not always generally known that mirrors are, simply, sheets of clear glass on to which a silver backing is sprayed, and which is, in turn, itself then backed with a protective layer of paint. So a mirror can, in fact, be treated by a leaded glass artist in exactly the same way as he uses other varieties of glass: it can be cut into different shapes as can normal glass and be wrapped in copper foil or housed in lead cames for incorporation into any glass project.

Clear glass and lead came combining to form an archway window inside the National Cathedral, Washington, DC. The design flows through the mullions and up into the traceries. Notice the various lead sizes
Robert Pinart, New York

Clear glass held together by soldered copper foil to form a geodesic terranium
Ingo Williams, Greenwich Village, New York

This adds another whole dimension on to the traditional use of glass, as pieces of mirror can be freely interposed with pieces of coloured glass within a design to greatly add to the range of artistic effects that can be achieved in glass.

In addition, the silvering process on a mirror is sometimes varied during manufacture to produce several single- or multi-coloured marbled patterns. (The only company that I know of that makes these is in Madrid (see Suppliers), but they do export, and these, likewise, can be most beautifully employed in a design.)

Combination of mirror and opalescent glass to form a bathroom mirror that is built into wood panelling and backlit so as to highlight the qualities of the opalescent glass
Alastair Duncan, London

Bullions

These are originally the centre of a sheet of spun crown glass. When glass is gathered, during manufacture, on the end of a glass-blower's pipe and spun around on its own axis, then the centrifugal force of the movement pulls out the blob till it forms a spinning disc.

When this disc has been annealed and broken off the glass-blower's pipe, then the juncture where the glass joined the pipe forms a thick, knotted knob of glass. This, known as a 'bullion', is nowadays specially cut out of the circular sheet of glass by the glassmaker and sold commercially.

Bullions are usually glazed at random with other varieties of glass, such as reamy antiques, to give a mediaeval atmosphere to public houses, Tudor hostels, and the like. They can, in addition, be very attractively employed in any glass project, but their charm lies in their being used with discretion. With bullions, a little goes a long, long, way. Do not cram as many of them as you can into a composition, as the effect is one of diminishing returns. Use them sparingly and at random as a decorative fillip to the rest of the glass.

Bullions are sold in various sizes – the largest being 16 in. × 12 in. and they can then be cut down to match required measurements. Remember to check the thickness of the glass to ensure that it is not too wide at any point for the lead came or copper foil that you are using, as it has an undulating border due to its being handblown.

Roundels

These are, in effect, bullions that are produced by the manufacturer in a finished, circular shape. They are made in America, and are spun and pressed in various colours, patterns, and diameter sizes.

Bullions do not have to be used in circular form. Here one is cut and incorporated into a window to form an eye-catching break in the symmetry of the diamond-shaped panels
Youngers Public House, Rathbone Place, London

Roundels wrapped in U-frame lead cames and soldered together to form a lampshade
Peter and Joanne Nervo, Berkeley, California

Glass jewels

These were widely used in Victorian and Edwardian door panels and fanlights, but seem to have fallen into disuse in England in recent years, to the detriment of the craft. In America, however, the ornamental charm of such jewels has made them, for the glass artist, of at least semi-precious value, and they are now pressed into a multitude of shapes, sizes and colours.

Jewels usually have a flat bottom surface and a raised top one, which is pressed into various shapes such as a convex, multi-faceted, or pyramidal form. This third dimension has a prismatic effect on the light rays, adding sparkle by both transmitted and reflected illumination. Whereas the centre of most jewels is raised, the width of the glass on the circumference is, however, usually approximately that of a sheet of hand-blown or machine-rolled glass, and so they can be wrapped in lead came or copper foil and incorporated into projects in exactly the same way as the latter. Like bullions, glass jewels are best used in a project to punctuate the overall beauty of the rest of the glass: their specific charm lies in their being used sparingly; too many jewels crammed into a glass composition tends to detract from the overall effect as they then no longer serve as adornments but rather as something that confuses and tires the eye.

A selection of glass jewels, some faceted, some not

Four globs; the one on the top left having an internal streaky pattern

Globs

These, too, are not an English tradition in glass, but they are considered an important and legitimate part of the range of decorative materials available to the glass exponent. They are humps of glass, varying in size, shape, and colour which are manufactured by heating small pieces of glass in a kiln till they melt into rounded 'globs'. You can, if you have access to a kiln, make your own, or, if not, purchase them commercially (see Suppliers).

Globs, like jewels, can be used to embellish a glass project by being leaded into the total composition. In addition, because of their having one flat surface (bottom), they can be glued on to the surface of sheet glass with clear epoxy adhesive to form such detail as eyes or dots. They provide a great deal of decorative effects to the imaginative artist, though they are less of use in the range of glass objects discussed in this book than to someone working in glass jewellery, figurines, and knick-knacks.

Glass imperfections

It is paradoxical that textural imperfections in glass, either on its surface or internally, tend to make it more, rather than less, attractive. They are, in fact, far from being the 'eyesores' that one might suppose. Flaws can arise, for example, through impurities getting into the glass compound during manufacture, but whatever their origin, they tend, on average, to increase rather than diminish the overall effect.

The reason for this is that any irregularity inside the glass will, on being struck by light rays, break them up into myriads of light particles that then refract in a multitude of directions. The effect can, at least, be eye-catching, and, at best, visually explosive. So always be on the lookout, when sorting through the racks at your glass supplier, for blemishes in the sheets of glass as these can be cut out – not to be thrown away – but to be used selectively within a design. Any piece of glass, however imperfect it is structurally, that is interesting, is not only a legitimate, but highly desirable, component of any work.

Detailed photograph of Tiffany's famous dragonfly lamp, showing his use of unfaceted oval glass jewels within the overall design. The dragonfly's wings are filigreed, patinated bronze superimposed over the glass

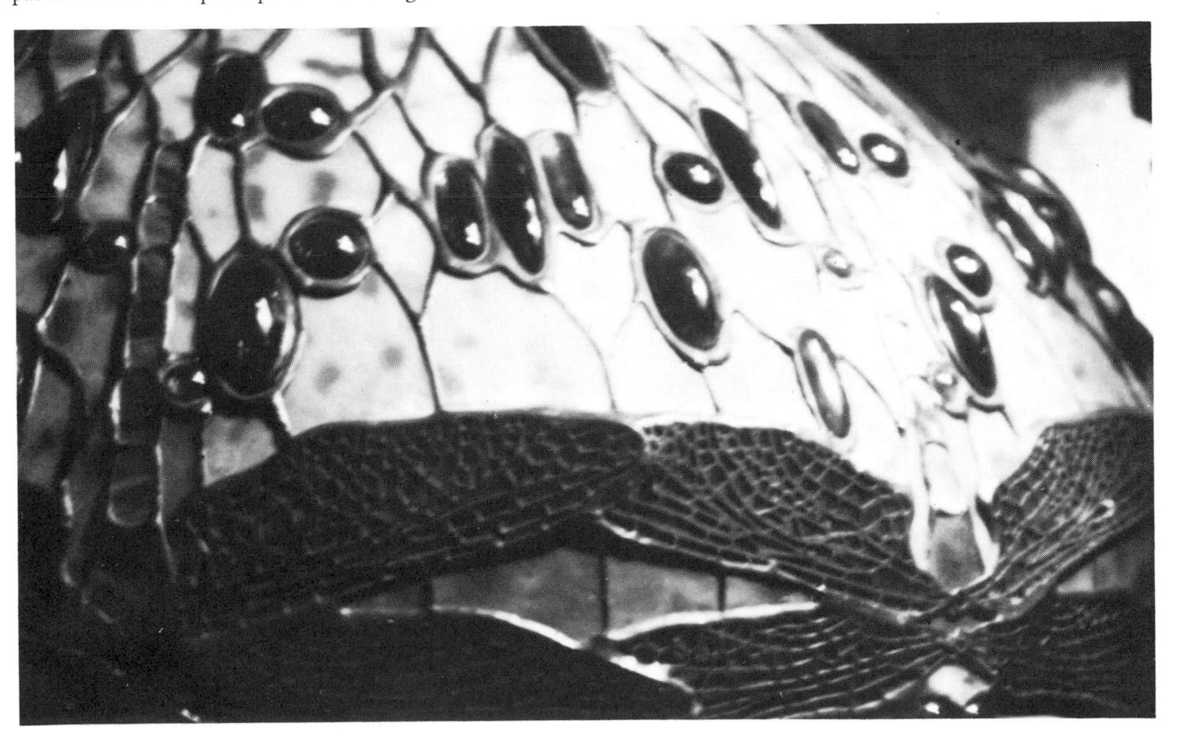

Summary

This chapter has discussed various sorts of glass available to the recently initiated glass enthusiast. Those mentioned do not, though, by any means constitute all of what is currently manufactured throughout the world, but they should give the newcomer a tip-of-the-iceberg awareness of the tremendous diversity of colours, shapes, and densities that glass as an art medium affords.

Almost any artistic effect can be reproduced in glass nowadays: there are, for example, varieties that are murky, limpid, neutral, sparkling, and blazing, which are just a few of the sensory impressions that the *right* piece of glass in the *right* composition can convey. The achievement of such effects is, however, up to the glass exponent himself: the effects are potentially there in the glass, and it is up to him to utilize them imaginatively and creatively. Ideas on how best to do this are offered in chapter 7, as there are several factors that must be taken into account when selecting the glass for a project. Remember, for example, that the colour of a single sheet of glass that your purchase from your glass supplier can appear radically different in hue and tone when it is cut into smaller pieces which are then placed in juxtaposition next to pieces of glass of other colours.

There is the additional factor that, while picking out the sheets of glass from your supplier's racks, you will probably see them partially by reflected light (this relates, especially, to selecting opalescent glass), whereas the glass will, finally, be viewed by transmitted light rays, once again producing a different effect to that which you perhaps envisaged when you bought it. Insight into colour harmony in glass does, in this instance, constitute 'throughsight'.

Experience will, of course, sharpen your colour sense and reduce the likelihood of glaring imbalances, and it is a good idea to develop the habit of setting aside a few minutes on the completion of a glass project to critically analyse how you could have improved, or otherwise altered, the finished colour scheme. There is no such thing as the optimum colour balance. There are always tonal nuances, however slight, that can improve every composition in glass, and it is important to continually work towards such betterment.

Another habit well worth developing is that of keeping an eye out for old, scrap glass, such as damaged Victorian or Edwardian window panes containing interesting glass, or offcuts of discontinued glass that a professional stained glass studio may let you purchase at a reduced price. Rummaging through antique shops can likewise bring to light a treasure trove of old lamps with broken panels. The prices of these are invariably marked down due to such breakage, and you can salvage whatever glass is reusable for your own projects. The wider the range of glass, in terms of hue, degree of opacity, and texture, that you the glass artist can acquire, the more diverse, in effect, is your palette.

Chapter 3

Tools of the trade

Work area

One of the marvellous aspects of leaded glass is that it requires so little in the way of ancillary equipment. All that is needed is a flat wooden work surface on which to assemble the glass and an electrical power point for the soldering iron. Obviously, the more space available the better, with ideally, racks for each colour or texture of glass, long boxes (placed horizontally) in which to store different sizes of lead came, drawers for tools and other materials, and ample room for a light table and a drawing board for the designs.

It is important to develop a disciplined method of working with glass as, more so than with most crafts, the materials themselves are inherently dangerous. Glass and lead, for example, should never be left within reach of young, exploring fingers and mouths. So try, at the end of each session, to put all tools and materials away and to clear all scraps of glass, lead, and solder from the work surface. Whenever possible use a vacuum cleaner as a brush just does not remove glass particles thoroughly enough, often leaving fragments that, although seemingly miniscule, will still draw blood. A neat, systematic work routine is, in addition, far more conducive to top quality workmanship than a work area that is littered with the aftermath of previous glass projects.

The following are factors that should be kept in mind when setting up and maintaining a work area:

1 Make certain that there is a power point within close proximity for the soldering iron. It will be necessary to switch the iron on and off at various time while soldering, so avoid working where you will have to stretch and bend to do so.

2 Keep all toxins, such as flux, patina, and blackening agents, stored well out of reach of children.

3 At the end of each session sweep the scraps of glass off the work surface into a cullet box. Before doing so, however, remove any pieces of glass that you feel you might be able to use in a future project. There is always something for which small pieces of glass can be used. Scraps can, for example, be assembled at random to form a crazy-paving abstract panel or cut into, say $\frac{1}{4}$ in. square tesserae for incorporation in a glass mosaic window or coffee table. In addition, glass is an expensive commodity that should be fully utilized, so always keep offcuts in the event of the need to replace broken sections in the project at some future time.

4 Try to buy and store sufficient supplies of materials to prevent running out midway through a job. There is little in life that is more annoying than to have to postpone completion of a project because of a miscalculation of how much lead came is required. This can be even more frustrating in the case of glass, as, on returning to the glass supplier to purchase the rest of the sheet from which the initial piece was chosen, you may find that someone has beaten you to it. So try to find enough space in the work area to accommodate reserves of stock. You are certain to use them eventually.

5 Make sure that the work area is well illuminated, preferably with white fluorescent tube lighting. Precision glass cutting and soldering requires it.

The following is a breakdown of the equipment required for the range of stained glass work covered in this book.

Glass cutters

These are described in detail because they quite literally 'make or break' the project. A cutter that functions properly is the very embodiment of the enjoyment that working with glass can bring. It scores the glass crisply, ensuring a clean and predictable separation of the glass, while minimizing glass wastage and time. But when it decides to be unco-operative then just the reverse occurs: one loses glass and time and, frequently, one's temper.

Diamond cutters

These are not suitable for our purposes. A great deal of the glass used in lamps, windows, etc, is not flat. Even the smoother, ie cutting, side of some glass is nodular or textured and a diamond cutter which has a very short point cannot always keep to the undulations so as to give a clean score, thereby leading to an incorrect break or a jagged, rough separation of the glass.

A batch of tungsten wheels
Reproduced by courtesy of A Shaw and Son, London

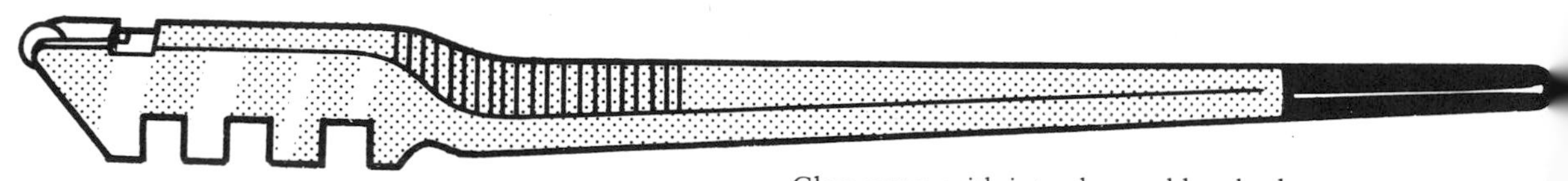

Glass cutter with interchangeable wheel

Method of replacing a wheel assembly
Reproduced by courtesy of Berlyne, Bailey and Co, Manchester

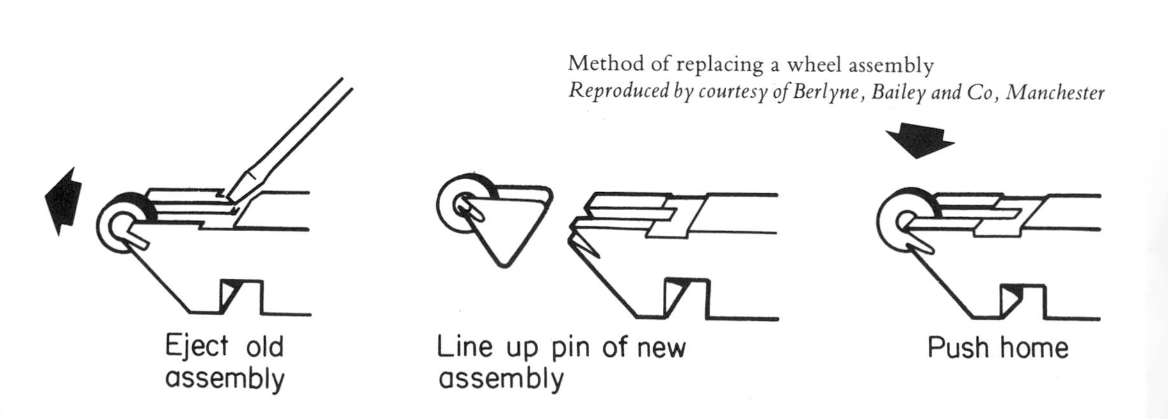

Wheel cutters

There are several excellent cutters readily available, for example, *Shaws* and *Diamantor*, while some glazing wholesalers stock the *Fletcher-Terry* and *Red Devil* makes from America. The final choice of cutter depends, of course, on the craftsman; but before settling on one try to handle as many different shapes as possible so as to determine which 'fits' you best. The following breakdown of the various components of a glass cutter is a guideline which, taken in conjunction with the different prices of cutters, may help you to make your choice.

(a) *The wheel*
This is either of steel or tungsten carbide.

Steel wheel cutters are general purpose cutters that retail for a modest price. The angle of the bevel on the wheel is about 122° which has proved to be the best shape for cutting a wide range of glass thicknesses satisfactorily. Steel wheels are, though, not long-serving and have to be replaced frequently as they become dull.

Tungsten carbide wheels are far more durable than the steel ones and are supplied with bevel angles ranging from 105° to 145°; the angle depending on the thickness of the glass and the hardness of the glass surface that is to be cut. The thicker the glass, the wider the bevel so that more pressure can be applied. The 145° bevel is, though, only supplied on wheels in an automatic cutting machine, and the angle on the tungsten cutters that is suitable for stained glass work is approximately the same as that on a steel one.

Tungsten carbide single-wheel cutters (without a ball-end) are a little more expensive than their steel counterparts, but are well worth the extra financial outlay. It is false economy to keep buying steel cutters when one has a great deal of cutting to do; rather invest in a tungsten cutter as it will, if well maintained, at least quadruple the service that a steel one can give.

Fixed and interchangeable wheels
Very few cutters are made with fixed wheels; most contain an interchangeable spare wheel assembly kit such as the one shown on page 34 which consists of a sheath holding the wheel on a pin (spindle). When the wheel has become dull then the whole assembly is ejected and replaced with a new one.

Interchangeable wheel kits are, of course, cheaper that having to buy a new cutter whenever the wheel becomes blunt, but make certain that the kit fits securely and, which is not necessarily the same, correctly, into the notch on the steel end of the cutter itself. If the notch is damaged in some way – such as being bent if the cutter is accidentally dropped – then the assembly may still fit into it, but the pin may not be at right angles to the body of the cutter. This, in turn, means that the wheel will be at an angle other than a right angle to the glass, and cannot, therefore, function properly.

Turret wheels
These must not be confused with the single interchangeable wheel assembly described above. They are round turrets that consist of two or six wheels as shown below. This design of glass cutter is not suited to our sort of glass cutting. The reason for this is that one cannot, because of the position of the turret, see the wheel as it scores the glass.

Turret cutter. Not for our purposes

(b) *Grozing teeth*
These are the three notches on the underside of most glass cutters. They are of different widths to allow for varying glass thicknesses, and their function is to snap off a piece of glass after it has been scored. The maximum width of glass that they can effectively break, however, is only about $\frac{1}{8}$ in. The notches are too narrow to generate a widespread, uniform pressure along a score line, with the result that the glass invariably chips or breaks incorrectly. They are, in addition, cumbersome and do not allow sufficient leverage.

Glass pliers are infinitely more practical for breaking pieces of glass off a sheet, and, certainly for our purposes, grozing teeth are an accessory that can be done without. I have never, in fact, seen them being used at all; and the fact that some recently marketed models of glass cutters do not include these teeth would suggest that they are now obsolete.

Using the grozing teeth on a cutter. Only practical for pieces of glass of about $\frac{1}{8}$ in. wide

(c) *The handle*
The handle, also called the *body* or *shaft*, of a wheel cutter is made of iron or wood. Practice will determine which you prefer. The iron handle is narrower than the wooden one, and some people find that because of this it has to be gripped more tightly to keep it steady, and that this, in turn, leads to chaffing between the first and middle fingers. The wooden handle is wider and less abrasive and can be more comfortable during prolonged cutting sessions.

(d) *The ball-end*
This is used to tap on the underside of a piece of glass directly under and along the score made by the cutter (see chapter 5). It is, I feel, an essential part of a glazier's equipment, though other instruments can, of course, improvise for it, but as the ball-end is designed specifically for the tapping function substitute tappers – such as the end of one of the handles on a pair of glass pliers – are unlikely to match its speed and efficiency.

Cutters that have ball-ends are considerably more expensive than those without them. It is not, of course, necessary to keep buying the ball-ended model. Only one ball-end is required and there are three alternative courses of action open:

1 To buy a ball-end cutter which has interchangeable wheels and to use this for all your cutting. This is not to be recommended, however, as spare wheel assemblies are finicky and temperamental gadgets that either get mislaid amongst all the stained glass kit on the work bench or that suddenly refuse to function properly. It is always a wise precautionary measure to have two or three spare cutters available for such contingencies.

2 To buy two wheel cutters; one with, and one without, a ball-end. Only one ball-end is needed so the initial purchase can be kept for all your tapping requirements, and cheaper, non-ball-ended cutters can be bought subsequently when the first two become blunt.

3 To buy a special-purpose glass tapper, for example, one made by *Shaws*, along with a cheaper model of cutter. This avoids inadvertently picking up the worn-out ball-ended one (which has been kept for tapping purposes only in 2 above) and then trying to use it to cut.

Price must always be a consideration in the choice of supplies, and is one of the factors (along with the quality and indispensability of the tool in question) that must be weighed in determining which cutter to buy. If you can afford it, I would recommend that you allow yourself the luxury of a top quality cutter and a soldering iron. Some tools can substitute for others and some, perhaps, can be done without, but these two are of the highest priority. So, in the case of the cutter, try to buy a good one. It will, in addition, if well maintained, be a cheaper investment in the long run than continually purchasing cutters of inferior quality.

Lubrication
Manufacturers of glass cutters recommend that a fine machine oil be used at regular intervals to ensure that the wheel can spin freely on its axle. Small particles of glass and grit get caught between the axle and the wheel and these prevent a smooth, continuous score being made; the wheel will skip or jam, which results in a broken score line.

To get maximum performance from a cutter it is a good idea to keep it in a jar of oil or kerosene when not in use. Rust can form on a wheel and this will corrode the cutting edge and leave it with an irregular circumference. This, of course, results in an only partially complete cut with each revolution of the wheel.

A lubricant does, in addition, facilitate the actual scoring of the glass itself. As a veneer of oil transfers from the wheel on to the glass during the scoring operation it works its way into the score and helps to keep the cleavage open. Glass molecules close up gradually after a score is made, and it is important to complete the fracture soon after cutting. Otherwise, if the glass is left for too long (a day, or for longer) then the break may not occur along the score; the molecules will have reformed, partly sealing the score, and, although the score itself is still visible, the break will not necessarily take place along it. Oil does help to prevent this.

Two glass cutters. One with, and one without, a ball-end.

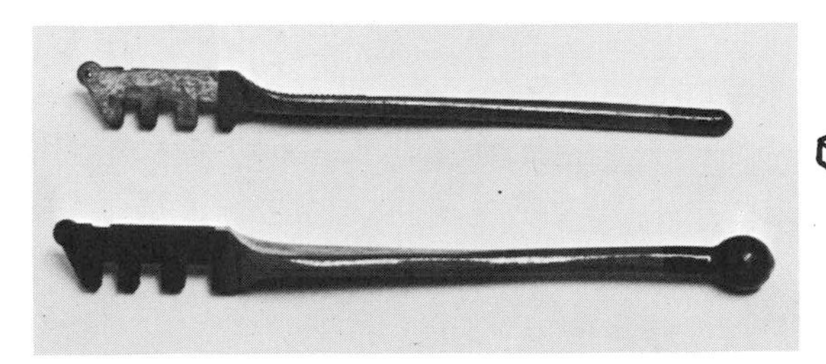

Circle cutters

Several makes of circle cutters are available, and these can be put to good use in stained glass. Their primary function is, as advertised, to cut circles (see page 96) – diameter sizes offered range up to 24 in. – but some models have a detachable cutting head that can be unscrewed and re-positioned for making straight cuts.

Most professional stained glass craftsmen would describe a circle cutter as being gimmicky; a superfluous, or at best, supplementary glazing tool that does not form part of a glazier's standard equipment. Beginners, however, often experience difficulty in cutting circles – the first attempts often result in an ellipse – because the pattern piece tends to slip as one shifts the glass (or, incorrectly, oneself) with each tangential cut made. A circle cutter alleviates this problem. It does, also, save time. Even when one has acquired sufficient proficiency in glass cutting to cut circles, output is greater with a circle cutter than with a pattern piece.

Glass tapper
Reproduced by courtesy of Berlyne, Bailey and Co, Manchester

Glass circle cutter, lens variety. The circle is inscribed on the glass by turning the top handle

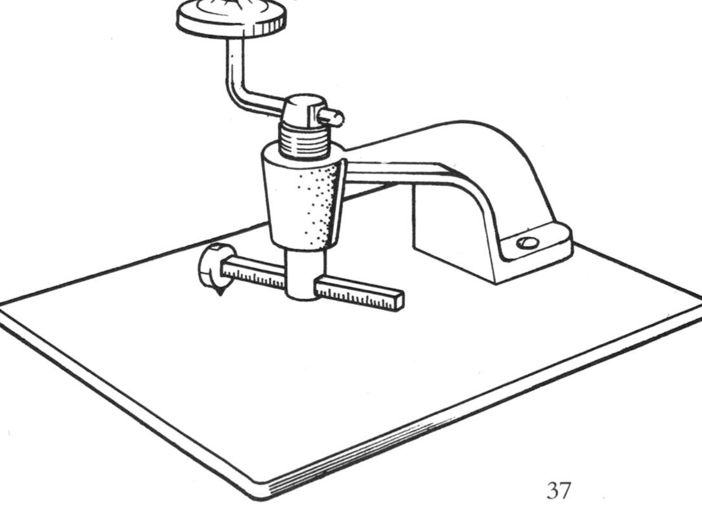

Light table

This is an ancillary piece of equipment that, although unnecessary for making the more simple range of leaded glass objects, increases in importance the more ambitious the designs become. You can, of course, make do without a light table by holding the piece(s) of glass against a light source. This, however, is a haphazard procedure, especially when it is advisable to compare more than two pieces of glass simultaneously for texture, colour, and striation (grain) effects. I am referring here specifically to the use of opalescent glass as its low translucency factor seems to 'trap' the light in the glass itself, thereby highlighting the actual textural formation of the glass. 'Streaky' glass, such as that available in England, does also have similar innate properties that, although lacking the same amount of opacity and incandescence, should still be checked against a light source before use.

A light table or box can be purchased from photographic suppliers, but these are expensive and often too large for purposes required. It is cheaper, and certainly not difficult, to make your own.

Make a square or rectangular frame out of five pieces of wood, leaving the top surface open. Fit two or three fluorescent tubes on the bottom, inside surface, and connect them to a single switch (ie not a switch per tube, as this is unnecessary) which should be situated on the outside of one of the sides of the frame. The box should be deep enough to prevent the heat of the tubes cracking the glass and also to be sure that there is room for the light to be sufficiently diffused.

Next, nail a narrow bead of wood around the inside of each of the sides, $\frac{1}{4}$ in. from the top of each side, and measure and cut a piece of glass (frosted, preferably) to fit inside the box on the ledge formed by the bead. Do not fasten the glass down as it will have to be taken out whenever a tube needs to be replaced.

You can, in addition, place a piece of silver of tin foil on the bottom surface of the box to reflect the light upwards.

A light table, such as this, is a useful piece of apparatus, and has a multitude of uses around the house other than for leaded glass work. It is also portable and can be stored out of the way when not needed.

Remember that the light table's primary function in the work at hand is to check colour patterns by juxtaposing pieces of glass of different hues to see whether they blend or not. It is not for cutting on as light tables in the conventional English method of using a cutline (see page 110). If, however, you wish to use it to cut on, then make certain that the glass that forms the surface of the box is thick enough to withstand the pressure generated in cutting.

A different model of circle cutter

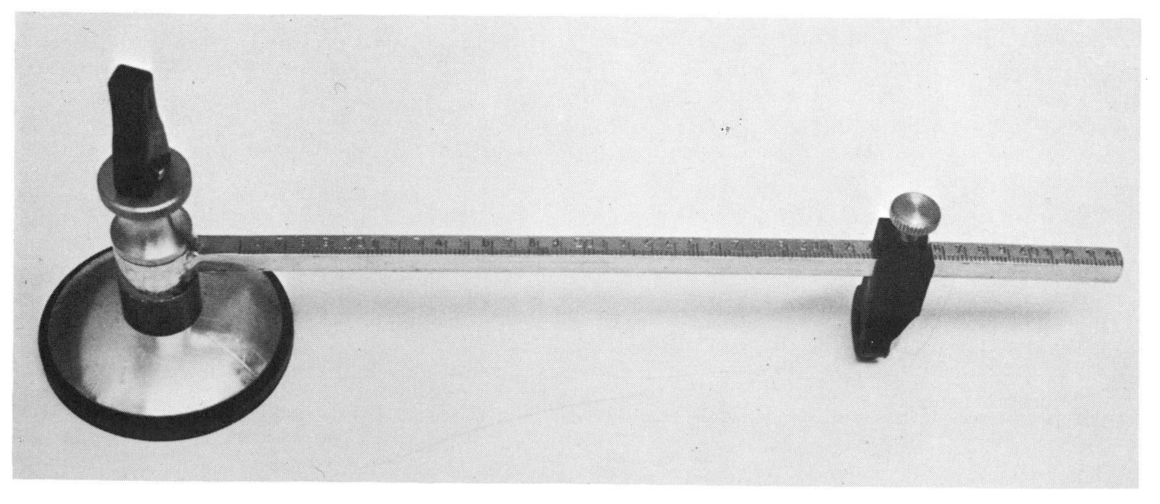

Work surface

This must be absolutely flat. Any unevenness will cause unequal pressure when scoring the glass, and the resulting break, in all likelihood, will not follow the score. This requirement is logical enough, but some glass, especially the antique variety, is often not entirely flat itself, so if it were to be cut on a flat surface then the pressure would likewise be uneven. To overcome the problem, place several sheets of paper (a newspaper is ideal) on the worktop. These sheets give the resilience necessary to absorb the pressures generated by any unevenness either in the glass or on the work surface.

Develop the habit of brushing off scraps of glass regularly with a brush. One is often tempted to sweep off accumulated glass particles and other debris with one's hand, especially late in the day when fatigue sets in, but believe me, as one who has on occasion resorted to such short-cut measures, the resulting cuts are often far from short.

Glass pliers

These, also known as glass nippers, come in various sizes, with those of lengths ranging from 6 in. to 10 in. and jaw widths from $\frac{1}{4}$ in. to 1 in. being most suitable for stained glass work.

The smaller the jaw the more easily the pliers can negotiate any narrow angles or sharp curves in the score line. Pliers with wider jaws are better suited to breaking off long strips of glass. The larger jaw area gives a more uniform pressure along the score line, and so increases the likelihood that the glass will break along the entire length of the score line and not snap off at some stage before the end.

As glass pliers all perform the same function, the choice of handle and jaw size is one of personal preference. Try, therefore, to handle various shapes and sizes before buying one.

Glass pliers are used to break off pieces of glass (they should be used instead of the notches on certain models of glass wheel cutters as they are a much more versatile instrument). They are of principal value where the piece of glass being scored is too small or oddly shaped to grip by hand. The pliers should be placed parallel to and as close as possible to (without touching) the score, and then pressure exerted firmly downwards to make the break. Nippers, unlike the grozers, should be used crisply so as to snap the glass along the score. Most nippers are, in addition, manufactured with grooves in their jaws, and these can be used as a substitute for grozers. Bear in mind, however, that the difference in the jaw design of nippers and grozers is because their respective functions are fundamentally different. If, though, the latter are unobtainable, then use the nippers instead.

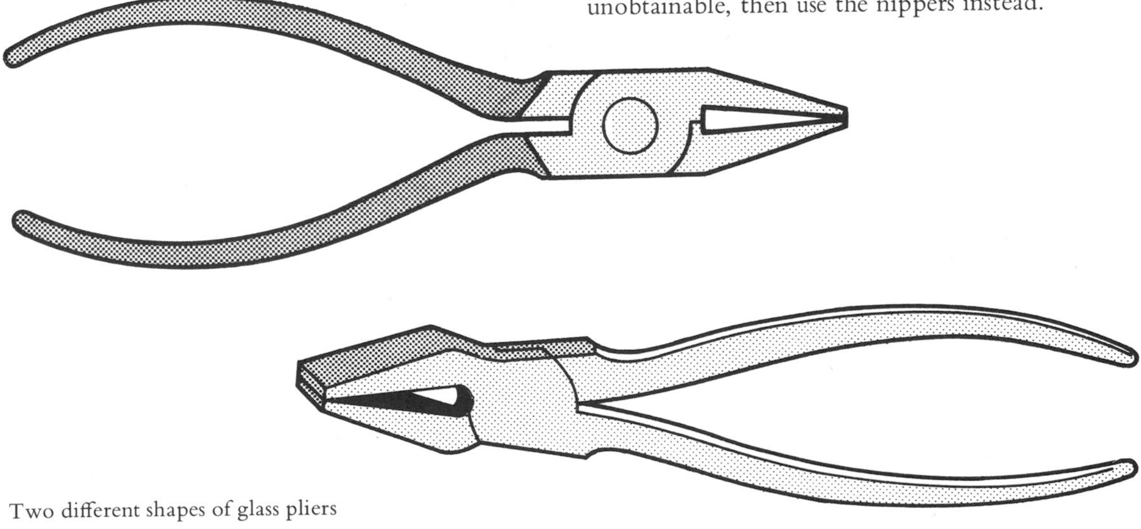

Two different shapes of glass pliers

Cut spreading pliers

These work on an entirely different principle to the standard glass pliers. They are used to 'squeeze' or 'run' a straight score line from the end of the sheet of glass, and not from the side as one would use a pair of glass pliers.

The jaws of the cut spreading pliers curve downwards: the upper one being concave and the lower one convex. The sheet of glass is positioned between the jaws so that its edge is at right angles to the pliers with the score aligned directly under the line marked in the centre of the upper jaw. When pressure is applied (gently, or the glass may crush) by squeezing the handles of the pliers together, the two ends of the upper jaw produce a downward pressure equi-distantly on both sides of the score, while the lower jaw acts as a fulcrum. The result is a clean, straight fracturing of the glass along the score.

Cutting long, thin strips of glass, such as might be required in panelled lampshades, can result in considerable glass wastage as the longer and thinner the strips, the more likelihood that the break will run off the score somewhere before the end. This can be prevented by placing a ruler under the glass (page 92) to give equal pressure along the entire length of the score, but cut spreading pliers are designed specifically for this purpose and do, consequently, produce the best results. Their high cost and limited use, however, categorize them as a non-essential, ancillary tool. A ruler can be used instead.

Using glass pliers to break off a long, thin strip of glass. Place the jaws of the pliers along the score-line and snap the glass with a downwards movement

Cut-spreading pliers

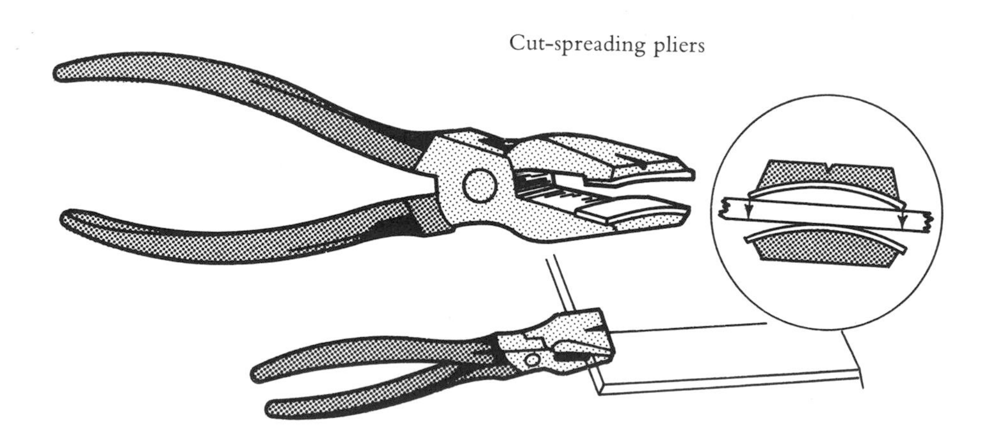

Offset pliers

These (also called *swan neck* or *rebent pliers*) are a variation on the standard glass pliers in that the jaws are designed specifically for cut-out and notch work. The hooked end of the jaws allows them to be held vertically to get into sharply angled curves where, due to confined space, there is not enough room to use standard glass pliers (which must be held horizontally to grip the edge of the glass).

A good example of the sort of glass-work for which offset pliers are designed is the removal of the inside of a circle cut in glass (page 96). Here the pliers are inserted into the hole from beneath the glass, and then used to break off the inside section.

Offset pliers are a specialized tool and offer limited benefits beyond those of the standard glass pliers. As such, they are an unnecessary purchase for most projects in stained glass work.

Grozers

These are flat-nosed, soft pliers that are used to 'groze', ie, to remove, any splinters or unwanted edges that remain on the glass after it has been scored and broken, or to eat away the glass in sharply angled curves. The jaws are serrated and narrower than those on glass pliers as its function entails a more delicate operation. Whereas the latter are used to snap pieces off a sheet of glass, the grozers are used gently to nibble or chew away unwanted glass in the same way that sandpaper would be used.

They are not always readily available at retailers that supply the glazing industry in England, and most glaziers improvise by making their own. This is done by heating the jaws of a suitable flat-nosed pair of pliers to take the temper out of the steel. When they have cooled, the edges of the jaws are rounded with a file.

Offset pliers. Few, if any, uses for leaded glass

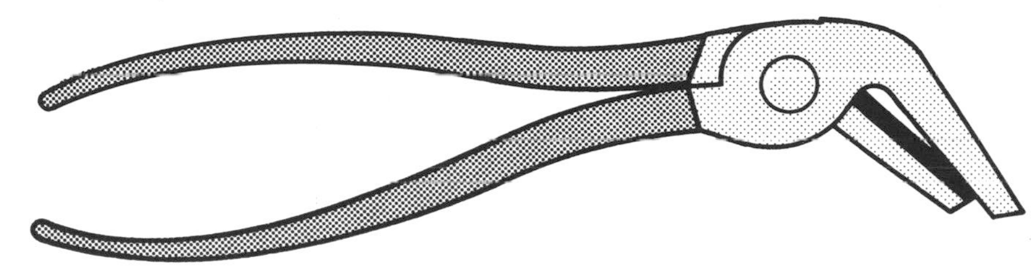

Grozers. Notice the parallel jaws

Slip-joint pliers

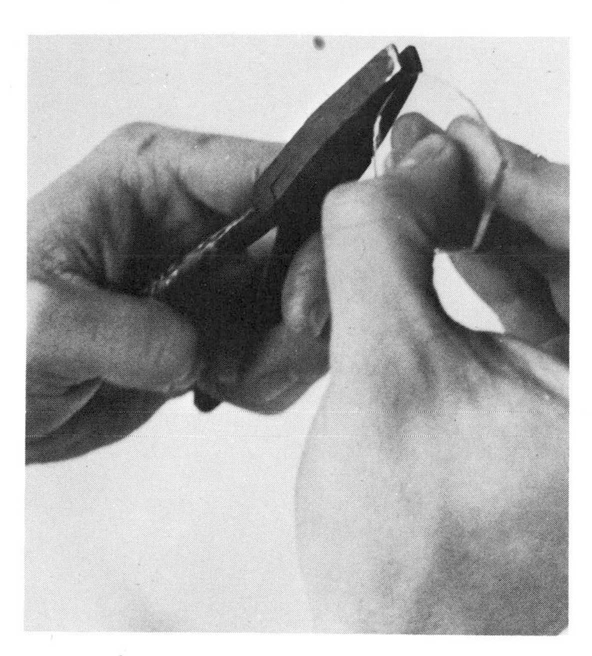

Using grozers to remove splinters from the edge of a piece of glass

A standard, all-purpose pair of slip-joint pliers should not be substituted for either the glass pliers or the grozers. Although the jaws are serrated and therefore allow a firm grip on the glass, they are not usually exactly parallel and invariably have some lateral movement which leads to a loss of control. Try to avoid them as they are not made for the job.

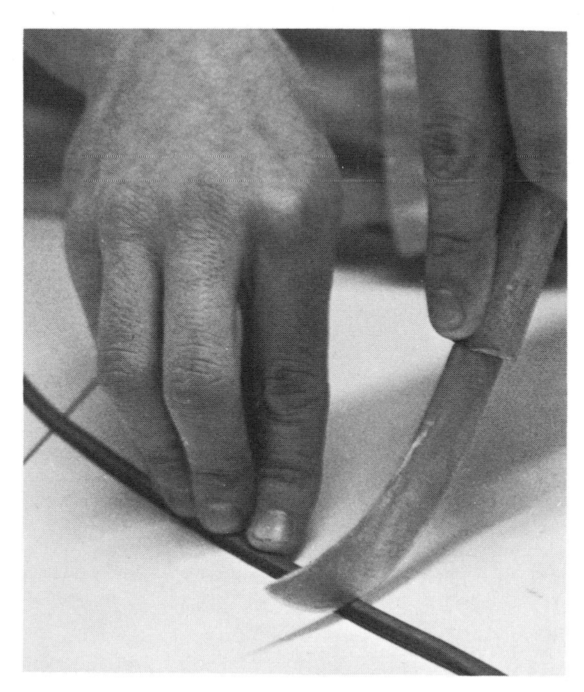

Cutting a strip of lead came with a lead knife

Lead knives

These can perform a multitude of functions in stained glass work, but are used primarily to cut and mitre lead cames. There is no standard lead knife as such, and the shape of its cutting edge, either curved or straight, depends on personal preference: a putty knife can be cut down and sharpened, or a household utility knife, or a linoleum knife, can be used. This last is the most suitable, but a lino knife has the cutting edge on the inside curve, whereas for cutting lead it is the outside edge that must be used. So either sharpen it yourself or ask the hardware dealer or ironmonger to do it for you. The end-product is a double-edged, dual-purpose knife resembling a pocket scimitar. Its hooked end can, for example, be used instead of a stopping knife, and it can be used to slip under a piece of glass to raise it to ensure that it rests on the lower leaf of the lead came into which it is to be fitted.

Professional stained glass craftsmen usually weight the end of the handle on their lead knives with solder and then square the end off so that it can be used as a hammer to tap a piece of glass so that it fits snugly into the channels in the lead came.

The best technique of cutting lead with a curved blade is to use a backward and forward rocking motion while exerting very gradual pressure downwards to sever the lead. Do not wield the knife as one would a meat cleaver; this will butcher the lead without cutting it and also blunt the blade. If, on the other hand, a knife with a square cutting edge is being used, then make certain that pressure is applied perpendicularly in a gradual, see-saw manner: too much pressure will cause the lead to collapse sideways, thereby crushing the heart and leaves. This, in turn, means that either you will have to open and re-straighten the lead end with a lathekin or pliers, or re-cut the lead.

Lead knives must be kept sharp, and this is easily done by honing them on a whetstone. Don't wait till your attempt to mitre the lead leaves it squashed or torn. Sharpen the knife at regular intervals by pushing the cutting edge along the stone as though shaving a thin layer off the stone. Keep the knife at the same angle throughout the stroke and then turn the blade over and repeat the process on the other side.

Two improvised lead knives. Above, a putty knife, and below a linoleum knife

Sharpening the edge of a lead knife on a whetstone

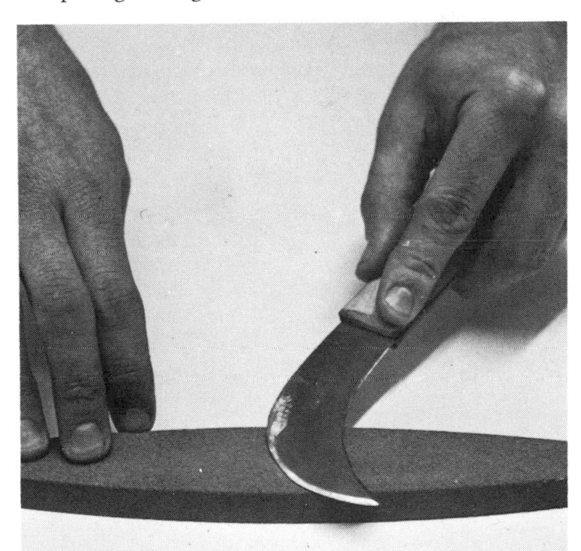

Lead vice

This is used to stretch lead came before use. Although a metal, lead is soft and malleable, and seems almost to have innate properties of elasticity. It is extruded during manufacture in long strips, and these must be stretched before use in stained glass work. The stretching process removes all the bends and kinks, hardens the lead, and increases the length of the strip. Place one end of the lead came in the lead vice, and then, gripping the other end with a pair of pliers, pull it by applying uniform pressure. The vice itself is not made for the stained glass profession *per se* – it is yet another improvization – but is used on yachts as a cleat to secure ropes. No matter; it serves our purposes just as adequately and can be purchased from any shop that stocks boating equipment. If one cannot be obtained, then lead can be stretched by two people gripping the lead at both ends with pliers, and then pulling it in 'tug-of-war' fashion. Alternatively, jam one end of the lead in a door or window frame and pull on the other end. Remember to apply pressure gradually as an over-ambitious jerk on the lead will cause it to snap suddenly, resulting, perhaps, in an unintended backward somersault.

Lead stretcher, or, to the yachting enthusiast, a boat cleat

Stopping knife

This is a standard glazing tool that, strangely enough, cannot be purchased as such in England, but must be made in makeshift fashion by the craftsman himself. It is used during the process of leading a glass project to ensure that the pieces of glass fit firmly into the channels in the lead came. During the process of building up a glass panel or window, just as one would assemble a jigsaw puzzle, there are always slight adjustments to be made in the positioning of the pieces of glass and lead came. When, for example, the heart of the lead extends beyond the line marked for it on the cartoon paper, then the blade end of the stopping knife is placed against the heart of the lead and gently tapped or levered to ensure that all empty space is taken up and that the lead is flush against the edge of the glass. A stopping knife is also needed to get into the lead intersections to prize open any crushed or damaged leaves on the lead before positioning the glass.

The proven, and thus preferred, shape for a stopping knife is one with a flat, tapered blade which is slightly curved at the end. The curvature ensures that any movement made by the stopping knife is horizontal, ie, parallel to the work top, so that any pressure that is applied is effective in moving the lead or glass sideways. The end of the blade itself must be rounded and blunt, because if it were sharp and pointed like a spike, then it would dig into the lead.

An oyster knife has the ideal shape for a stopping knife, and all that is required is to curve the end of its blade. This is easily done by holding it over a flame and then, when it is red-hot, bending it with some household pliers. The end of the handle is often squared off with solder to allow it to be used as a hammer.

An improvised stopping knife; an oyster knife with upturned head

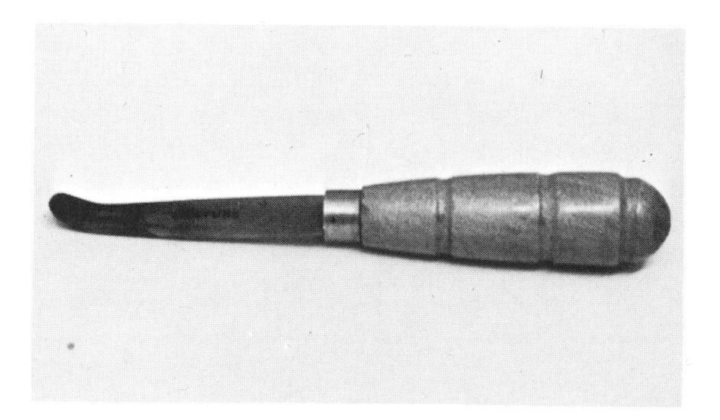

Lathekin

A lathekin is used to open and straighten the channels in a strip of lead came. If the lead has been stretched prior to using, as it should have been, then the use of a lathekin is minimized; all large kinks will have been removed during stretching, and the lathekin need only be run along the channel perfunctorily to ensure that it is flat and straight, and that the leaves of the lead are not bent or buckled.

Lathekins must be made by the craftsman himself in England as they are not sold commercially. A flat piece of wood or bone is suitable, and can be shaped, for example, like the one illustrated which was part of a ruler prior to being whittled down. The edges of the tapered end should be smooth so that the lathekin does not catch in the channel as it is manoeuvred along it. It makes no difference whether the lathekin is pulled or pushed. Do whichever is found to be more effective – but make sure that the lead is flat and straight before you begin. Lead that is correctly stored, ie, in long strips, is far less likely to need attention than lead that is bent into coils for storage and must consequently be untwirled.

Of all the words in the glass craftsman's lexicon, 'lathekin' is the one that most beginners find hardest to remember. It may, perhaps, function just as satisfactorily as a 'thingummy' or 'thingammyjig', but, as in any profession, the correct terminology is a kind of shibboleth that distinguishes the person who is conversant with his work from the one who is not. So try to get it right from the start.

A piece of ruler whittled down to form a lathekin

Using a lathekin to open the channel in a lead came

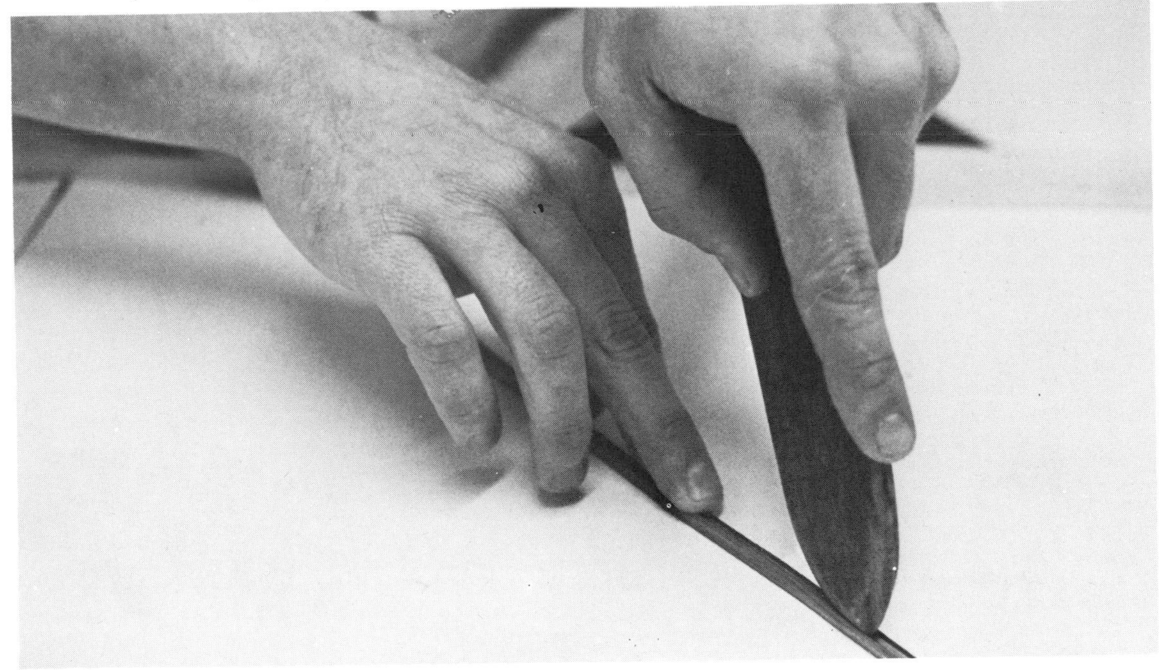

Hammer

This is required to knock nails into the wooden work bench while working on a project in order to secure the lead and glass while building up a glass panel, and to nail strips of wood in position to act as a frame for the panel itself.

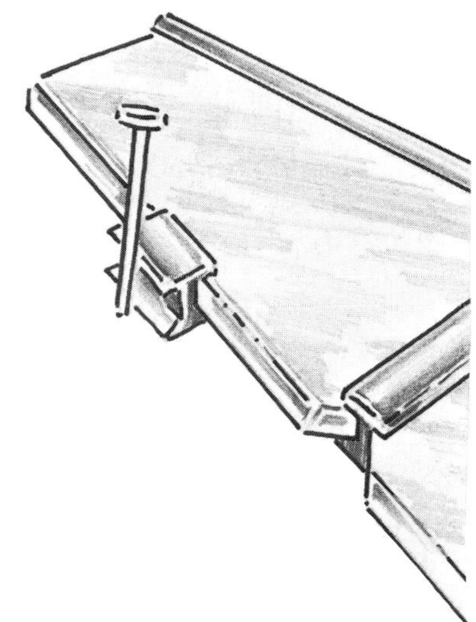

When using standard, circular nails it is best to place an offcut of lead came between the nail and the glass to prevent the edge of the glass from chipping

Hammer with plastic and thermoset rubber heads

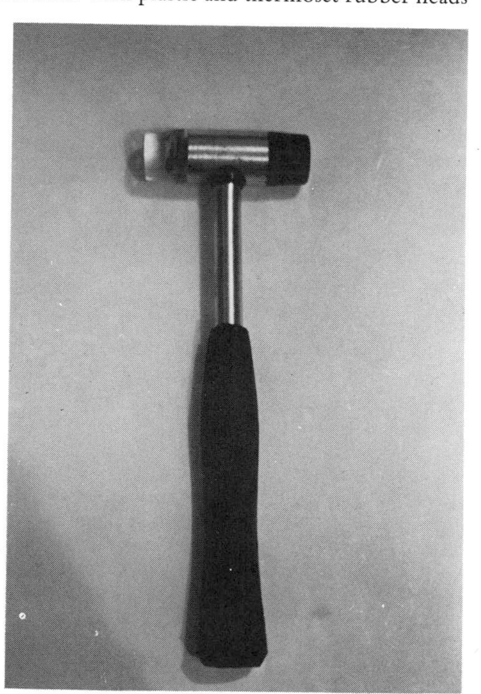

Nails

Their function is to hold the lead came and glass in position while the stained glass project is being assembled on the work table. Some nails, however, are more suitable than others. Due to the fact that they are placed flush against the glass or lead came, ideally they should be tapered and right-angled. A standard nail is circular and can, while being knocked into the work top, create an abrasive force that chips the edge of the glass, or in the case of the lead came, is likely to crush the top leaf of the lead. Horseshoe nails (known also as farrier's nails) are perfect for the job. However, when it comes to trying to obtain them they can be most elusive. Ironmongers do not normally stock them, and one is usually directed to go to the local blacksmith, who, in turn, is nowadays as elusive as the nails themselves! For this reason, I should not over-emphasize their importance, but do try to get hold of a dozen or so rectangular nails. Standard nails will do the job, but be careful when knocking them in. They should be about 1 in. long with a small head. They should not have a large, circular head like that on a drawing pin, as this tends to catch on one's fingers as one continually uses and re-uses them in building up the glass panel.

A good method, when using standard (circular) nails, of preventing them from chipping the edge of the glass, is to use an offcut of lead came between the glass and the nail. Small chips on the edge of a piece of glass will be hidden under the leaf of the lead came in the completed project, and will therefore not be detectable. But larger ones will extend beyond the lead, and then the piece of glass should be re-cut. To go to all the trouble of re-cutting the piece because of a small chip that is barely visible may seem to be unnecessarily meticulous but it will permanently mar the work.

Two sizes of horseshoe nails

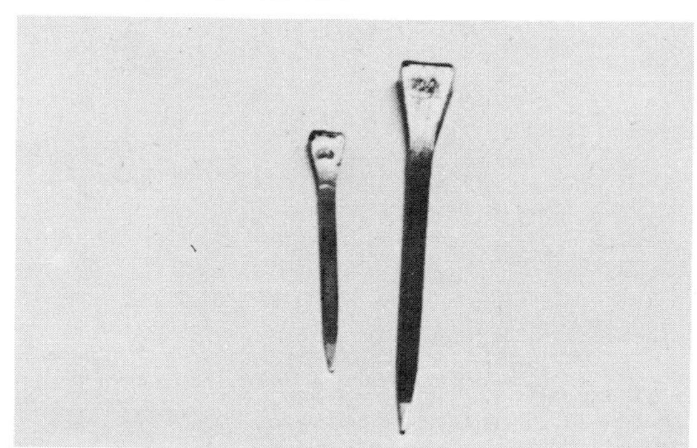

Moulds

These are necessary if a lampshade cannot be made up in flat panel sections and then assembled by bending and/or soldering the sections together. Lamps that are rectangular or polygonal can be made flat on the work bench and then bent to reach their intended shape (see chapter 9). Shapes that are curvilinear, however, usually require a mould; especially for the traditional Tiffany-styled lamp designs such as the hemisphere, dome, cupola, or tulip. Whereas such shapes can, in certain instances, be built up with lead came without a mould, they simply cannot be managed satisfactorily when one is using copper foil (see chapter 11).

Moulds can be made of virtually any material: those stocked by art and crafts suppliers in America are often of plaster of paris or styrofoam, though commercial craftsmen usually use wooden moulds as they are long-serving and do not lose their shape or tear as do moulds made of materials such as plaster of paris or styrofoam or even plastic. The combination of the heat generated in soldering, liquid fluxes that have a dampening and corrosive effect on the surface of the mould, and even the accumulated weight of the glass as it is assembled, all lead to wear-and-tear. Unless a batch of, say, five or more lamps of the same shape are being made, then plastic or plaster of paris (which can be replastered) ones are adequate as wooden moulds are expensive.

It is a good idea to keep an eye out for objects that can improvise as moulds; there are probably several spherically-shaped bowls in every kitchen that would, when turned upside-down, make attractive lampshade shapes.

A selection of cardboard lampshade moulds

Two wooden moulds; one cupola-shaped, the other tulip-shaped

Dual line cutter

This consists of an adjustable two-blade assembly that can be set to $\frac{1}{16}$ in. to cut the width of paper for the heart of the lead came when removing pattern pieces from pattern paper.

The blades remain parallel whatever the width, thus ensuring an accurate end-result.

Pattern shears

These are specially manufactured to cut pattern pieces out of pattern paper so that the exact amount of space is left between the pieces for the heart of the lead came. Pattern shears consist of three blades: two on the bottom and one on top. The bottom two have a $\frac{1}{16}$ in. gap between them into which the top blade fits during cutting, thereby removing a $\frac{1}{16}$ in. strip of paper.

The shears can be used with either the two blades or the single one uppermost; if using the former technique made certain that the line being cut falls between the two blades; if the latter method is employed, then ensure that the single blade falls directly over the line as you cut along it. Either method is suitable, though short cutting strokes with the back of the blades should be used, or else the strip of paper being removed tends to clog up the shears.

Pattern shears are expensive and therefore do not warrant such a large financial outlay when the makeshift cutter (page 00) performs at least as well.

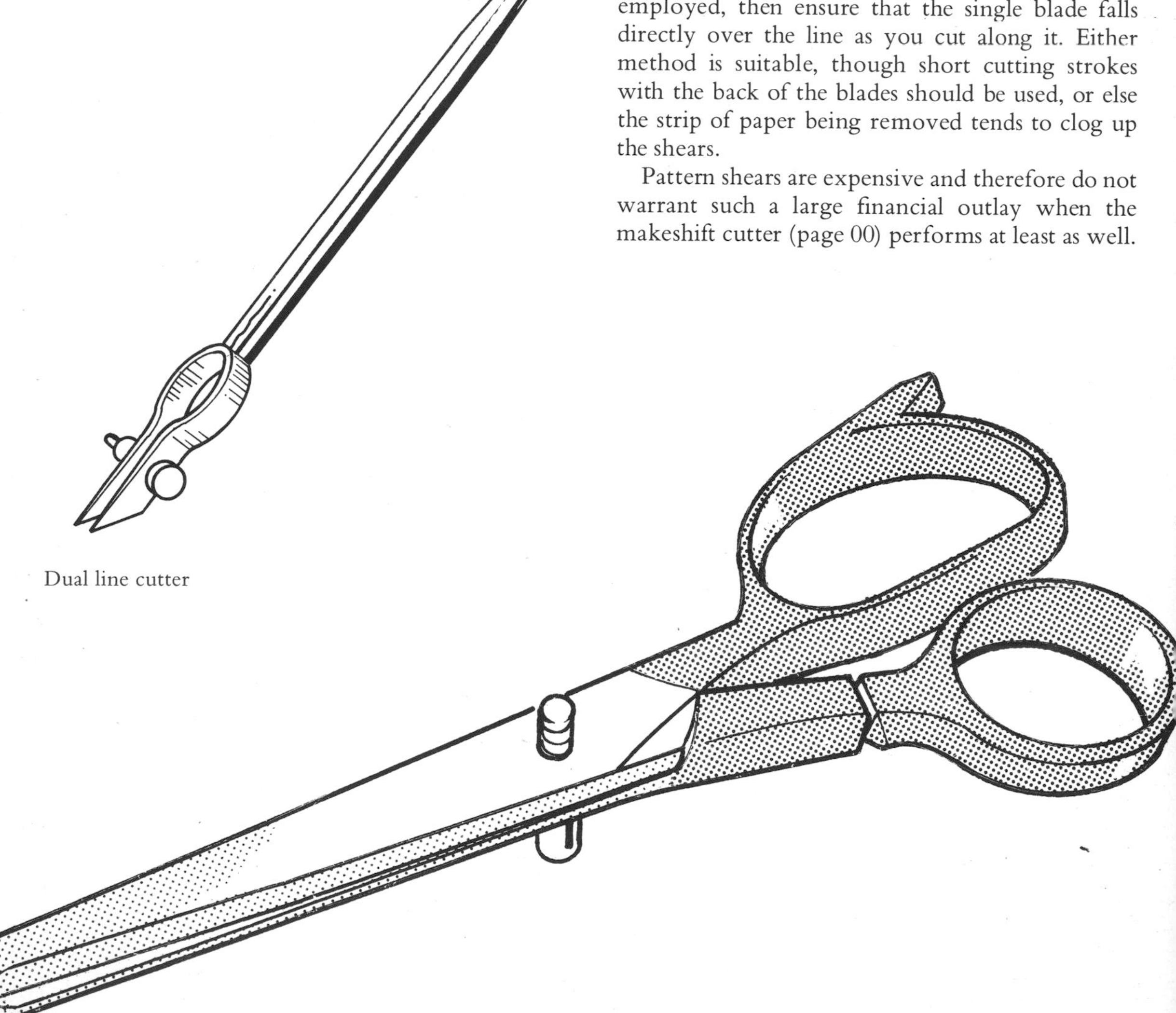

Dual line cutter

Pattern shears

Triptych window
Ray Bradley, London.

Art Nouveau mirror
Alastair Duncan, London.

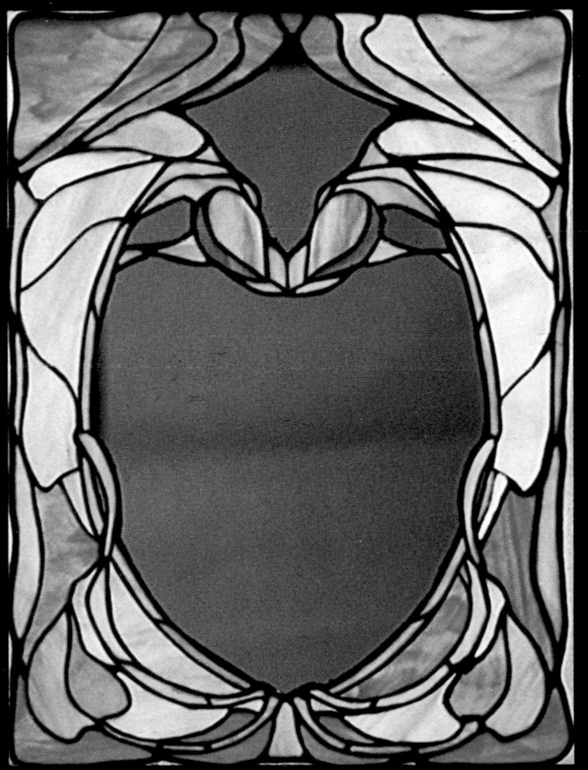

Johann Sebastian Bach
Jeff Speeth, Friendship New York.

Table lamp

Pattern piece cutter

This is used to cut the pattern pieces out of the pattern paper described on this page so that a gap of $\frac{1}{16}$ in. is removed from between each piece for the heart of the lead came that borders on each piece of glass.

Neither pattern shears nor dual line cutters, both described on page 48 are readily available in England, but a functional and cheap substitute is to take two single-edged razor blades, place a $\frac{1}{16}$ in. piece of cardboard or wood between them, and then bind them together with adhesive tape. It performs accurately and with adequate manoeuvrability in cutting around sharp corners.

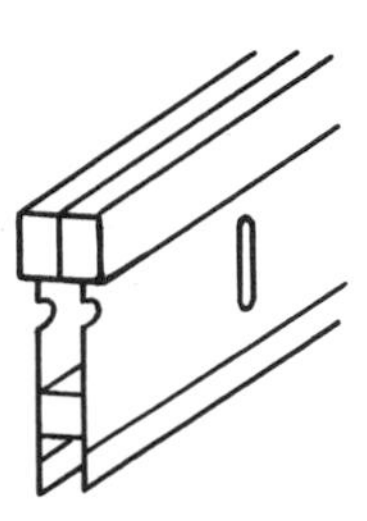

Two single edge razor blades

Papers

Glass projects such as panels and windows require three copies – a cartoon, a pattern paper, and a work drawing – of the original design, while others, such as lampshades, often only need the pattern paper. The following makes of paper, for each respective function, are recommended:

1 Almost any make of paper is suitable for the cartoon, although it should not be too thick as the design on its has to be transferred by means of carbon paper on to the work drawing and pattern paper (see page 105). A heavy white cartridge is suitable.

2 Pattern paper must be of a much stronger texture – about the thickness of a manilla folder – as it has to withstand the wear-and-tear of the wheel of the cutter as it is drawn along the edge of the pattern piece. It must not be too thin as then it will buckle or become corrugated. If, however, it is too thick then problems arise if the distance between the axle and the wheel is less than the thickness of the paper. In this case the wheel will not reach, and therefore not be able to score, the surface of the glass.

3 *Kraft* paper, such as the standard brown paper used to wrap parcels, is suitable for the work drawing. It will be placed on the work bench and needs to be fairly vigorous to withstand the effect of nails, hot solder, liquid flux, and other abrasive action, as the glass project is assembled on top of it.

4 In addition to these papers, it is advisable to keep a supply of finely grained sand paper. This is of use when the project requires very thin leading (such as the $\frac{3}{16}$ in. flat or round H-leads available in England). In such instances the glass must be cut with meticulous care to ensure that there are no chips along the edge of the glass surface (these would be hidden beneath a wider leaf, but they will, in this instance, show beyond the edge of the leaf). The thinner the lead, the narrower the leaf that overlaps the glass, and so for the thinnest lead widths precision rounding off of the edge of the glass is required so that the glass and lead fit flush against each other without showing any gaps or chips in the glass.

The use of sand paper, although far more time-consuming than grozers or nippers, prevents the edge of the glass from chipping off incorrectly during grozing.

Brushes

Stained glass work entails the use of at least three differently textured brushes:

1 *Flux brush*

This is used to rub the flux paste or liquid into a joint after it has been de-oxidised prior to soldering. The brush should be of a fairly hard bristle (hair, not wire), such as a cut-off paint brush, so that the flux can be worked right into the newly milled surface of the joint. The flux acts as a catalyst between the solder and the lead came or copper foil in the soldering process, and the more thoroughly it covers the area to be soldered, the better the bonding of the solder. Do not worry about excess flux remaining on, or staining, the glass around the joints. The flux used in stained glass is not corrosive and washes off easily with soap and water, so apply it over-generously rather than sparingly.

2 *Wire brush*

This is used to remove the oxidization that forms on lead came. The lead at each joint must be burnished prior to soldering to remove the oxide that has formed on the lead; if the lead surface is not cleaned then the solder will not bond itself readily to the lead, causing messy or incomplete joints. Use a soft wire brush if available, or, if not, then steel wool, a razor blade, or knife, can be used. (The small copper brushes sold by shoe shops for suede shoes are ideal.) Do not use force: undue pressure can scratch the glass or rip the lead came itself. A light, brushing motion will remove the layer of oxide, leaving a bright, milled surface to the lead underneath. A wire brush used for this purpose must not become a multi-purpose instrument: do not use it for fluxing or cleaning the work bench.

3 *Bench brush*

It is wise to develop the habit of brushing the scrap bits of glass off the work surface into a container otherwise they can so easily attach themselves to one's person and cause cuts on the hands and to clothing. It is a good idea, also, to groze over a bin as this will catch the fine splinters of glass. The work surface must, in any event, be brushed frequently to ensure that it is completely flat and clean.

Lamp-making stand

This is an ingenious device consisting of a stand positioner, adjustable arm, and a set of interchangeable discs to facilitate the assembly of glass lampshades on moulds.

The normal procedure of building up a glass lampshade on a mould entails placing the mould on the work bench and, starting from the centre at the top, adding each piece of glass in clockwise or anticlockwise fashion around the mould. This ensures that, as more and more pieces are added, the weight of these assembled pieces is equally distributed, ie, balanced, around the centre of the mould. As, however, one progresses lower and lower down the mould, adding each new piece of glass, so it is necessary to tilt the mould so that each additional piece of glass can be fitted horizontally in order to 'tack' (lightly solder) it in position. The lower one goes, the more pieces of glass and, of course, the greater the accumulated weight. This, in turn, means that the mould and glass combined becomes increasingly unwieldy as one props it up each time to tack on a new piece of glass, and this generates both wear-and-tear on the mould – especially if it is made of a light material such as plastic or plaster of paris – and continual problems of glass slippage as one props up the mould each time a new piece has to be soldered.

The lamp-making stand eliminates these difficulties. The kit contains a series of discs that act as supports for different sizes of moulds. The mould is placed on top of the disc and then fastened to the threaded arm with a nut. The centre of the mould must, of course, be hollow, such as plaster of paris and plastic moulds, which only need to have a hole pierced in the top so that they can be placed on the arm. If, however, the mould is solid, then a hole will have to be drilled right through it so that it can be fitted.

The lamp-making stand can adjust to any position so that the area to be soldered remains level at all times. The stand positioner allows the arm to rotate and tilt to any angle, thus eliminating all the problems of propping up the mould on the work bench.

Lamp-making stand with interchangeable discs

Lamp-making stand with a mould attached

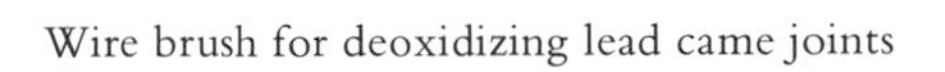

Wire brush for deoxidizing lead came joints

Lead finishing agents

(a) *Blackening agent* This is a special paste or liquid solution that is rubbed on to the surface of either soldered lead cames or copper foil to give it an antique appearance. The special blackener darkens the leading immediately, giving it a permanent grey or black finish (the depth of colour depends on how much is applied). Certain intermediate steps are, however, necessary in both cases.

Lead darkens naturally as oxidation takes place (more quickly out-of-doors, where it is exposed to the elements, than indoors), but this is a gradual process, and the resulting hue may not be as dark as is required. The blackening agent will not, however, directly darken the lead so a layer of solder must be applied to it. First, de-oxidize the lead's surface by milling it with a wire brush or steel wool, then flux and apply a thin layer of solder. When all the lead is covered smoothly by the solder then wash off all remaining flux and incidental dirt, and rub in the blackener with a cloth or paper towel, being as careful as possible to prevent it getting on to the surface of the glass as if this is pitted, then the blackener will work its way into such holes, leaving dark spots in the surface of the glass which are often impossible to completely remove.

When using copper foil the process is much simpler as the foil is bonded together by solder, and so all that is required is that the blackener is rubbed on directly after soldering. Remember, again, to wash away all flux and dirt before blackening.

The method of application varies with the various brands marketed, though it is usually rubbed on to the solder and left for half an hour or so before being scrubbed off with soap and water. The result is a uniformly darkened skeleton of leading around the pieces of glass that greatly enhances the overall effect, while removing the project's new, somewhat cheap, appearance.

An ideal blackener for our purposes is the 'Gun Blue' paste sold by gunsmiths for darkening the barrels of shotguns. Any gunsmith stocks it, and the instructions for application are on the label.

(b) *Patina* Copper staining powder is available from chemists in the form of copper sulphate crystals. This gives a bronze patina finish to the soldered lead or copper foil. (The method of preparing the lead is the same as for the blackening agent, ie, it must first have a layer of solder applied to it.)

Certain colours of glass, such as the green, yellow, and ochre range, usually show up better when bordered in leading that is patinated than that which is blackened. When applying the copper crystals ($CuSO_4$), dampen them with warm water and rub the solution on to the solder with a cloth (the depth of colour depends on the strength of the solution and the number of applications), and then leave it to dry for a half hour or so before scrubbing it with warm water and cleaning powder.

Glass scriber

This is used for writing on glass and has a tungsten carbide point with which to score the surface of the glass. Its use is described in chapter 12.

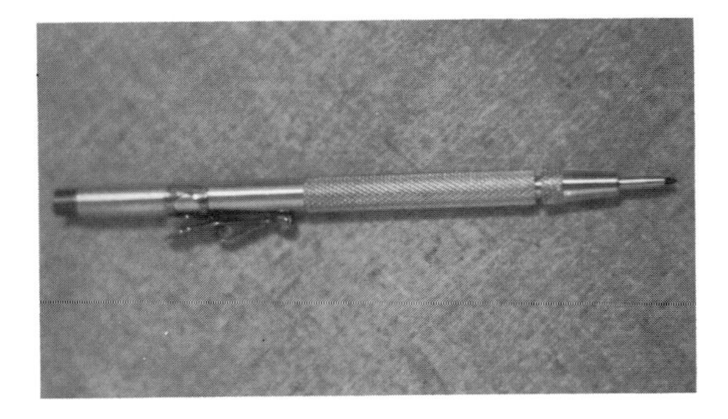

Safety glasses

Safety glasses are not always readily available and are not necessary if glasses are normally worn. A pair of sunglasses or lenses of a very low strength can be used as a substitute, the only prerequisite being that there is a protective shield in front of the eyes to intercept any flying glass particles. Glasses can be a nuisance, but the inconvenience is nothing compared to the damage which can be caused by a sharp splinter of glass lodged in the eye.

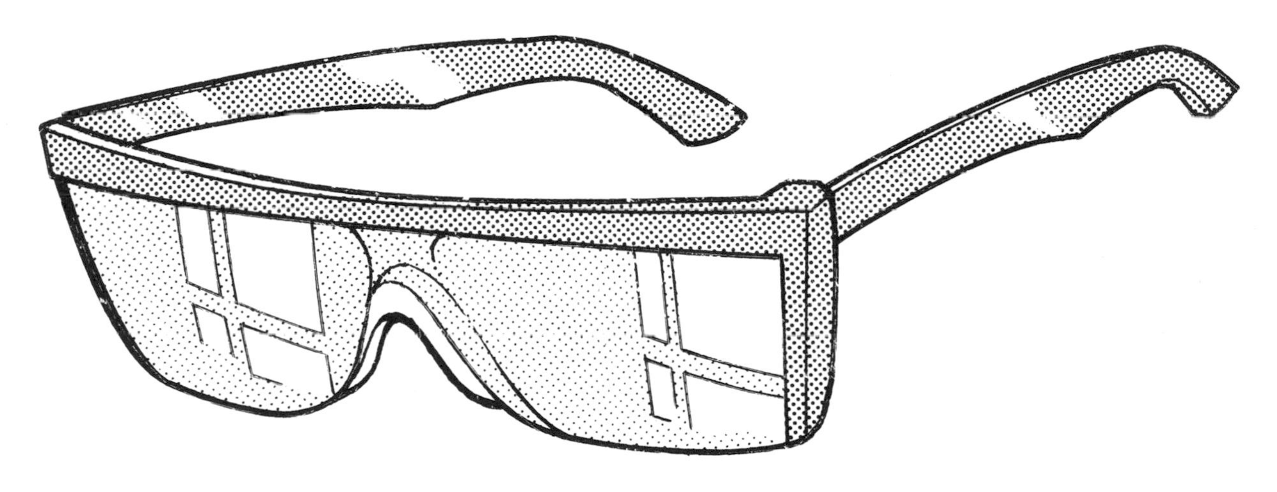

Safety glasses

First aid kit

Keep a tin of sticking plaster within the immediate proximity of the work area. However proficient at glass cutting you are, there are occasions (diminishing in number as you familiarize yourself with glass-handling techniques) when your fingers will get cut by glass splinters. These, for the most part, will be small nicks or pinpricks, but it is amazing how much blood such a small incision can generate. The only consolation is that glass is so sharp that the cut is usually painless.

As soon as you find that you are bleeding stop what you are doing, wash your hands, and make sure that the piece of glass is not still in your finger. Then put a piece of plaster over the cut.

The temptation is to finish 'just one more section' before doing this, or to lick or wipe your finger to remove the blood, hoping, in the meantime, that the cut will stop bleeding of its own accord. Even a pinprick cut on the tip of a finger can cause the blood to spurt out, and, unless the flow is stopped immediately, your finger becomes a sort of 'tar-baby', transferring blood on to everything it touches. In no time at all the whole work area becomes messy and sticky, and this is not, of course, conducive to good craftsmanship.

But most important is the aspect of hygiene. Dirt accumulates in an open wound and, unless removed, will become infected. This refers not only to a splinter of glass that, if not removed, tends to work its way into one's skin, but also to working with lead came. Lead, as any printer knows, is, if allowed to get under the skin, injurious to one's health, and can cause lead poisoning. There is, in addition, the possibility, if one works with dirty, cut fingers, that blood poisoning or tetanus can occur.

The above may seem alarmist, but I do feel the importance of being earnest about safety measures in working with glass and lead. Always cover cuts and always scrub your hands thoroughly when you have finished a work session. Further references to hygiene in glass handling will be made in later chapters so as to reinforce the need for due care, and I make no apologies if I tend to stress this point repeatedly. Anyone who has undergone anti-tetanus or lead-poisoning treatment will appreciate the need for precautionary measures.

Left
Different sizes of lead came stored in long, horizontal boxes

Right
Glass racks showing neat, systematic filing of pieces by colour and opacity

Chapter 4

Lead, copper foil and the soldering process

This chapter describes the components that are used to hold the glass together in glazing. Pieces of glass cannot adhere to each other, and so a means has to be found whereby they are held in position: a means that not only locks them securely and permanently together, but one that will increase, or at least, not detract from, the overall aestheticism of the project. This function is performed primarily by lead came or copper foil (the latter is described in chapter 10), which are held in place by soldering.

The chapter is divided up into five main sections:

1 Lead came
2 Soldering irons
3 Solder
4 Flux
5 The soldering process.

The reason for this rather exhaustive breakdown and description of the materials and technique required, is: firstly because I am assuming that the reader will be starting from scratch and will, therefore, require a detailed analysis of all the whys and wherefores, and the dos and don'ts, of this aspect of leaded glass work. There is not as of now, as far as I can ascertain, any printed information available in England that covers the entire range of materials and techniques of lead came, copper foil, and soldering. Specific (and excellent) information exists on certain equipment and procedures, but no literature on the subject exists *in toto* for the initiate.

Secondly, this dearth of information has led to a general lack of knowledge of the range and functions of the materials in question. My attempt to remedy this situation in this chapter is due, in fact, in no small measure to my own frustrations in trying to discover exactly what – both chemically and mechanically – actually occurs during the soldering operation. Formal instruction in glazing is not widespread, and the only other real source of information is the professional craftsman himself. And he, being a member of what is now an alarmingly diminished trade, is, in turn, less and less accessible to supply the necessary tutelage.

Thirdly, the projects in glass described in this book are all intended to be articles of virtue and, as such, must be able to withstand any critical examination of the quality of workmanship involved.

The distinction between the windows in chapter 8 and those made in the professional stained glass trade may serve to show why it is perhaps even more important to be concerned with the quality of our leading and soldering than it is for professional glaziers whose work is often placed high above the viewer's head in an apse, transept, or clerestory.

The windows described in this book, however, are intended to be used primarily for decorative purposes in the home – they are non-professional, non-traditional, non-church windows which will be displayed so that they can be clearly seen (ie, not only seen *through*). They will be able to be critically scrutinized, not only for their artistic merits, but also for their workmanship. It is important to realize that the more visually accessible the project, the better finished it must be.

Most professionally made windows (especially, again, church or cathedral windows depicting biblical or other religious scenes) contain glass that is painted on to give the necessary detail. One's eyes concentrate on the painted areas, and the leading and soldering fade into the background to become of secondary importance; the lead came being present largely for its function in holding the painted glass together. The glass windows described in this book, however, do not have painted designs on them, and so the leading should play a far greater role as an integral part of the composition as a whole, not only to divide up areas of glass for creative effect, but also as a work of art in itself. Neat, untorn lead cames and inconspicuous soldering are niceties that bring refinement to the finished object. In the same way that a badly chosen or damaged picture frame will ruin a beautiful painting, so untidy leading and soldering will mar the total effect of the glass that they surround.

This, if anything, applies even more to lamps and other three-dimensional glass objects than to windows. Lamps can be picked up, turned over, and examined from any angle. The soldering on the inside of a lampshade (especially with copper foil) is a tell-tale method of determining the overall quality of the workmanship involved. Many a lamp that, on first glance, seems the epitome of perfection, will, on being turned over, show an unfinished or shoddily soldered inside. So, in order to silence any would-be critic, or to ensure the marketability of the lamp, or simply for the self-satisfaction of knowing that you have achieved a certain degree of virtuosity in your work, pay close attention to the soldering technique.

The technique of soldering, more than other aspect of leaded glass work, is something that you may need to refer back to as you seek to acquire more and more finesse in your projects. At first the soldered joint will be of secondary importance: the appearance of the solder will be largely inconsequential – as long as it 'sticks' the lead together. The magic of your first creations in glass will be that they exist in their entirety, and not as a composition of individual parts of the whole. The amateurish and unfinished appearance of any badly soldered joints will probably go unnoticed, or be subconsciously repressed, as you bask in the glow of your first glass lampshade.

It is usually only with experience – as you become confident in handling the materials and techniques of the glass craft – that critical analysis of your own, and others', workmanship will develop. Suddenly that first project will, on reappraisal, no longer meet your ever-increasing standard of technical acceptability. Large humped and messy joints, which before seemed adequate, will now appear as mountainous blemishes that stick out like the proverbial sore thumb, and you will probably be amazed that you could have been satisfied with them previously. This is, as in any new undertaking, only natural, and is a very good indicator that your ability for qualitative appraisal is becoming more and more finely attuned to the medium – glass and lead – with which you are working.

The identification of a problem and its cure need not, however, necessarily be one and the same. That a soldered joint is unsightly is one thing; how to correct it may be something entirely different. Simply going over it again with the soldering iron will probably not alleviate the problem: if it did not form a well-rounded and tidy joint at the first attempt, then it may not do so in subsequent efforts. There are four ingredients that go into the formation of a neat joint – the lead came, the soldering iron, the solder, and the flux – and if any one, or a combination, of these is not functioning properly, then this will undermine the whole operation, no matter how many times you attempt to remedy it.

It is at this stage, therefore, that it may be necessary to refer back to this chapter to determine why that mutinous lump of solder is behaving in such an impeachable manner and is driving you to the very threshold of sanity. Perhaps the soldering iron is dirty or too hot, or perhaps you are using the wrong flux. If the reason why the soldering operation is malfunctioning is not immediately identifiable, then you will have to analyse each component – soldering iron, lead came, flux, and solder – in turn to determine whether it is deficient in any way. The solution to the problem will, hopefully, be found in the following pages.

Casting and milling lead in the eighteenth century
Le Viel 1774

Lead came

Strips of lead are called *came* or *calme* in the stained glass profession (the word derives from Old English, and means 'strips'). Their function is to provide structural support for the glass: to interlock all the pieces so that they hold firmly and permanently together. Lead cames are the traditional method of binding glass together, having been used, in conjunction with glass, in the construction of the church windows that first cast light on congregations in the Dark Ages. They still perform their function admirably, although there are recent innovations (such as copper foil) that can substitute for lead came.

The characteristics of lead are unique for a metal of such a high specific gravity. It has both rigidity and flexibility, tensile strength and malleability. When it is soldered into place around the glass it has the structural properties of iron, yet when being shaped to follow the contours of the glass or being stretched prior to use, it has those of rubber or *Plasticene*. This combination of strength and pliancy is ideal for leaded glass.

Lead came is formed from the raw lead by one of two processes: either by milling or by extrusion. The former involves an electrically-, or rarely, a hand-driven machine in which the lead is drawn through a vice by two wheels (like a mangle), which grip the inside of the heart of the lead and pull it through a pair of split dies. These shape and cut the lead into its H-shape form. The latter is a process of drawing the lead through fixed dies by hydraulic force.

Although the properties of the lead came produced by the above two processes are slightly different, this is of no consequence here. What is important is the range of shapes and sizes of lead came that are available commercially, whether milled or extruded.

H-frame lead came
These are 5 ft lengths of lead made in a variety of shapes and sizes (in America the length is 6 ft); the main two being the H-flat and the H-round ones.

H-frame leads: their use and misuse
The design aspects of H-frame leading will be covered more extensively in chapter 7, but it is important to bear in mind, as you familiarize yourself with the various widths of H-frame lead cames, that leads are a primary, and *not* a secondary, part of any design. They are not there solely to hold the glass in position, but also to complement it artistically. By judiciously mixing the widths of the leads in the window, a multitude of different creative effects can be achieved. So do not be 'mis-lead' into assuming that they are incorporated solely for functional purposes.

This is not to suggest that lead widths should be varied indiscriminately: obviously such a mélange will have a negative impact, both confusing and tiring the eye.

Remember that there are two potentially limiting factors to the use of narrow and wide lead cames: first, whereas narrow leads (such as the $\frac{3}{16}$ in.) are supple and wispishly unobtrusive due to their thinness, so, for this same reason, they do not supply much strength to the glass. Be careful, therefore, in designing windows that are entirely, or in large part, constructed with narrow leads, as these will be structurally weak (unless the problem has been anticipated at the design stage by carefully balancing the leading).

Wide leads, conversely, give the required rigidity, but, in doing so, forefeit a large amount of flexibility. Leads that are $\frac{1}{2}$ in. and wider cannot be bent to follow the contours of an acutely angled piece of glass, so there are constraints on their implementation in intricate or angular glass patterns.

Left
Cross-sections of just some of the over 100 different lead came designs manufactured by the White Metal Rolling and Stamping Corporation, Brooklyn, New York

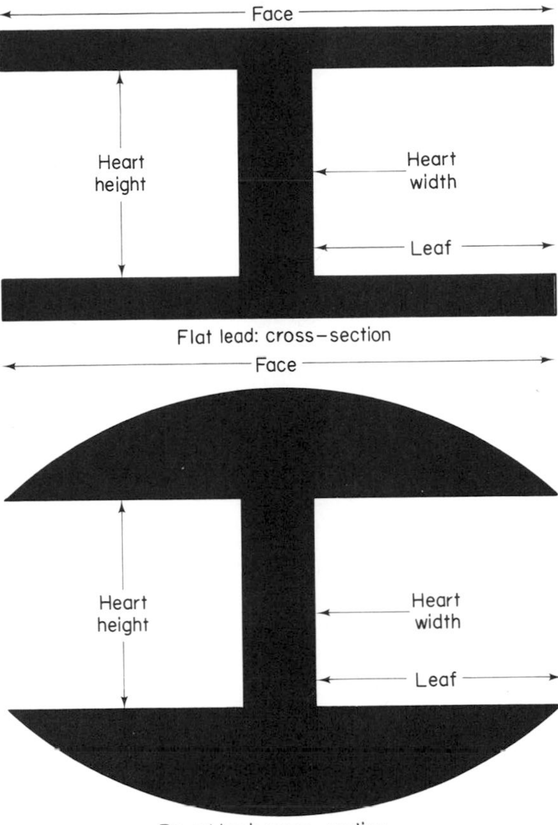

The dimensions of H-frame flat and round lead cames

Dimensions of H-leads
H-frame leads are shaped like an H placed on its side, ⌶, and have the following three dimensions:

1 *Face width*
The main measurement given to describe the size of a strip of lead came is the width of the top surface, called the face. All references to lead size in this book and in ordering from your supplier, unless specified, relate to this measurement. When, therefore, you order a $\frac{1}{4}$ in. lead, the stockist will understand that you want a lead that has a width across its face of $\frac{1}{4}$ in. It is only when you require a strip of lead came that has a special heart height (see below) that it is necessary to give an additional measurement.

2 *Heart width*
This relates to the width of the lead that forms the vertical section of the ⌶: the central partition that separates the two channels into which the pieces of glass fit.

The heart width is $\frac{1}{16}$ in.[1] and not a measurement that needs to be quoted in ordering lead came, because all lead, whatever its face size, has the same heart width. It is important, however, always to allow for the $\frac{1}{16}$ in. when using pattern pieces (see chapter 6), because if you do not, then the overall dimensions of the panel will be correspondingly too large when the glass is assembled.

3 *Heart height*

This is the height of the two channels in the lead came: the distance from the inside of the top leaf to the inside of the bottom one, ie, the space into which the pieces of glass are fitted.

The different sizes of heart heights available in England are: $\frac{1}{8}$ in., $\frac{3}{16}$ in., $\frac{7}{32}$ in., $\frac{1}{4}$ in., $\frac{5}{16}$ in. and $\frac{3}{8}$ in.

The standard size – the height of channel which is wide enough to fit around most varieties of machine-rolled and hand-blown glass – is the $\frac{3}{16}$ in. Lead came is, unless specified, normally supplied with a $\frac{3}{16}$ in. heart height, so unless you are using glass of an abnormally large or small thickness, it is only necessary to specify the face width when ordering lead came.

Before choosing lead, however, it is important to measure the thickness of the sheets of glass from which the pieces in the panel will be cut. Determine the width of the thickest section along the edge of the glass before selecting the dimensions of the lead came that you will use. Some hand-blown glass, such as certain antique English and Flemish varieties, vary considerably along their edges, and unless allowance is made for the thickest part, it will not fit into the channel in the came.

Few things in life test one's patience more than to be half-way through a project and then to discover that a piece of glass will not fit into the channels. And no amount of widening of these with a lathekin or stopping knife will solve the problem. The normal sequence of events is to first try to wiggle the glass into position, then to tap it gently with the handle of the lead knife, then to tap it harder, and harder. This usually ends up as a major exercise in frustration. All to no avail. Not only will the leaves of the lead be irreparably concertinaed, but you will, in addition, probably chip or break the glass.

You will either have to replace the glass (probably with a piece of another colour, which will disrupt the intended colour balance of your composition) or change the lead for one with a larger heart height. Both operations are time-consuming and could have been prevented by a preparatory check that the height of the heart is sufficiently wide to house the entire piece of glass.

H-frame leads: available sizes

Both H-flat and H-round lead cames are available in England in the following face widths: $\frac{3}{16}$ in., $\frac{1}{4}$ in., $\frac{5}{16}$ in., $\frac{3}{8}$ in., $\frac{7}{16}$ in., $\frac{1}{2}$ in., $\frac{5}{8}$ in., $\frac{3}{4}$ in., and $\frac{7}{8}$ in. while the selection in America is infinitely larger.

The English range is generally adequate for our needs, but if there are occasions when you may wish to order a specific H-frame lead from America, then I suggest that you plan well in advance so that it can be delivered by surface mail. It is not just the weight, but also the cost, of lead that can make a hole in your pocket, especially if it involves delivery by airmail.

H-leads are primarily inside leads as they are designed to fit on the inside of a panel, window, or lamp so that both of its channels contain pieces of glass. The only time that an H-frame should be used as an outside lead on a project is when it is used to allow for a margin of error, on the large side, in fitting a window into a predetermined space. In such instances the outer channel of the perimeter leads will remain empty to be filed down or bent inwards if necessary, but as it will finally be enclosed by the moulding that secures the window in position, so it will not be visible.

Any project that uses lead came and that has an outer edge that is visible must, for aesthetic reasons, use a U-frame. An empty outer H-frame channel gives an unfinished, non-professional appearance which – although this can be remedied somewhat by shaving off the outer channel and filing down any remaining roughness – will still be apparent. A place for everything and everything in its place: H-frame leads are inside leads.

[1] The width actually varies between $\frac{1}{16}$ in. and U in. but is generally taken to be $\frac{1}{16}$ in.

Flat or round H-leads

The choice of whether to use a flat or round H-frame lead in a particular circumstance is one of personal preference. The flat variety tend, in practice, to be used as perimeter leads, while the round ones – which, because of their bellied shape contain more lead – tend to be used as inside leading as they are structurally stronger. In addition, the rounded leads are better suited to lamps which contain panels of glass that are bent into shape (see chapter 9), because the rounded face of the lead came blends in more harmoniously with the overall shape of the lamp than does the flat lead.

There are, though, no hard and fast rules, except that once you have decided on one or the other shape within a project, then do not vary the two at random as this is artistically incongruous.

Lead stretching and glazing in the eighteenth century
Le Viel 1774

Sizes of H-leads

$\frac{3}{16}$ in.

The narrowest face width of lead in England is the the $\frac{3}{16}$ in. (a $\frac{1}{8}$ in. is, though, available in America). This requires precision glass cutting as the leaves of the lead came do not overlap the surface of the glass to any great extent, thereby allowing no margin of error in your cutting. Any chip in the glass will show beyond the leaf of the lead came, while any incorrectly cut piece of glass will result in a gap between it and the adjacent piece. The $\frac{3}{16}$ in. therefore requires a high degree of accuracy in glass cutting (this is, incidentally, in no way intended to imply that larger sizes of lead came require less exactness), and so it is a good idea not to incorporate it in your projects until your cutting expertise has developed somewhat. Small inaccuracies along the edge of the glass or chips on its surface can be sandpapered down (see page 49), but if you are a new-found enthusiast embarking on a first project, then the precision required for $\frac{3}{16}$ in. work may have an overall disheartening effect. So begin, therefore, with the $\frac{1}{4}$ in. or $\frac{5}{16}$ in.

$\frac{1}{4}$ in. and $\frac{5}{16}$ in.

These are widely used in both panels and lamps, and a large percentage of your lamp work will probably incorporate one or other of these two sizes as they are wide enough to grip most pieces of angled glass, while not being too heavy to overshadow the glass. Most people find that the $\frac{1}{4}$ in. lead becomes their general-purpose lead – providing sufficient rigidity to hold the glass firmly in place, yet also being unobtrusively narrow. A word of caution, though: whereas it does perform both of these functions, I would suggest that you do not allow yourself to become too conditioned to it, as this can only inhibit the creative effect that varying your lead sizes can produce. (The design aspect of varying lead widths is discussed in chapter 7.) Leading that is all one size is at best monotonous, at worst artistically debilitating.

$\frac{3}{8}$ in., $\frac{7}{16}$ in. and $\frac{1}{2}$ in.

These are, in all likelihood, the widest lead cames that you will use as a home craftsman – being used selectively, ie, sparingly, in windows and panels, while forming the frames for large-sized lamps. They are best used linearly as the wider the lead the less flexible and supple it is for use in curvilinear designs.

$\frac{5}{8}$ in., $\frac{3}{4}$ in. and $\frac{7}{8}$ in.

These are, almost without exception, outside the range of our needs, being too large for most lamp work and used only on very large windows – more for structural support than for artistic effect. They are used primarily in professionally-made stained glass church windows where the dimensions of the window require that it has a strong frame. In such instances the incorporation of steel core leads (below) or steel bars, in addition to, for example, a $\frac{7}{8}$ in. lead frame, is often necessary.

Half-rounded beaded lead

This is available in the following sizes: $\frac{7}{32}$ in., $\frac{1}{4}$ in., $\frac{5}{16}$ in., $\frac{3}{8}$ in., $\frac{1}{2}$ in. and $\frac{5}{8}$ in.

It is a compromise between the flat and round H-leads: whereas the bellied shape of the round lead gives it additional strength, the flat lead is sometimes considered to be more decorative. The half-rounded beaded lead, therefore, is a half-way measure: combining the ornamental appearance of the flat lead with 75% of the tensile strength of the rounded one. It is not a commonly used lead largely because glass craftsmen are often unaware of its existence. It deserves more attention, though, as it brings an additional dimension to the lead came. Use discretion, however, as sizes above $\frac{5}{16}$ in. can be visually overpowering. A little goes a long way.

Unequal flat leads

These are H-lead cames in which the heart is off-centre, so that one channel is deeper than the other. They are only available in England in sizes of $\frac{1}{2}$ in. and $\frac{5}{8}$ in., being only infrequently used in very specialized leading operations.

There are two main uses for it. The first is where the design in your window necessitates the structural strength of a $\frac{1}{2}$ in. lead, but which requires a less than normal leaf width on one side of the lead, because the leaf on a standard $\frac{1}{2}$ in. H-lead would swallow up too much of the glass for the effect that you want to achieve.

The second use for unequal flat leads is where one of the pieces of glass fitting into the lead is marginally too thick for the specific channel heart height of the lead, and therefore needs the additional width of the wider channel to help grip it in position.

You are unlikely to use unequal flat leads very frequently, if at all, in your projects, though you should still be aware of their existence.

Steel core leads

These are lengths of lead came that have a hollow heart into which a steel rod is inserted to reinforce their overall strength.

The lead came is supplied in 5 ft strips with the steel already inside it, and the came must then be cut by the craftsman to the required length with a hacksaw. They are used to strengthen large windows and should be incorporated at about 24 in. intervals into any window that is larger than approximately 3 feet square (for the design aspects and constraints of steel core leads, see chapter 7).

When using such 'hole-in-the-heart' lead cames, remember to make allowance for their heart width – which is larger than the average $\frac{1}{16}$ in. of standard leads – on your cartoon. Failure to do this will result in the dimensions of the completed window being larger than you had intended; an oversight which is as easy to make as it is difficult to rectify once you have completed your soldering.

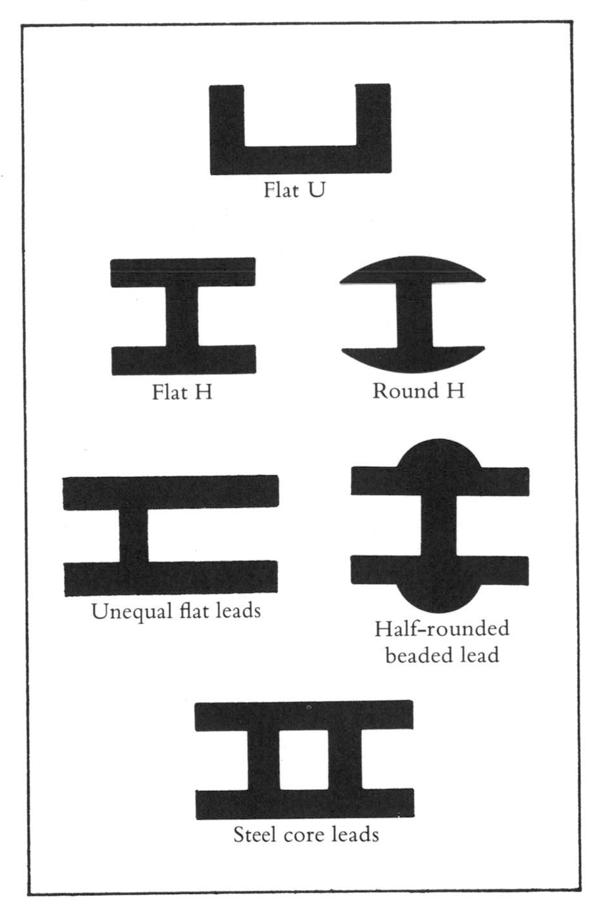

U-frame lead cames

These have only recently been manufactured in England, and are now available in sizes of $\frac{1}{16}$ and $\frac{3}{16}$ in.

U-frames, unlike H-frame lead came, have only one channel and are, therefore, outside leads, ie, they are used on the borders of glass projects such as the perimeter of a panel or window or on the skirt of a lampshade. They form an irreplaceable part of leading equipment, especially in the construction of any three-dimensional glass objects that have an exposed border, where the use of an H-frame lead would look shoddy.

$\frac{1}{16}$ in.

The $\frac{1}{16}$ in. measurement refers to the leaf of the lead (ie not to its heart height). It is, therefore, a very narrow lead that offers very little strength; its purpose being to give the glass a decorative finished appearance.

It is best used on machine-rolled glass as this is of a standard thickness, whereas hand-blown antique glass has an undulating edge which cannot effectively be gripped by a lead leaf that is only $\frac{1}{16}$ in. wide. The $\frac{1}{16}$ in. U-frame lead is widely used in the United States by home craftsmen for making free-form objects such as mobiles and general glass knick-knacks, which, when taken in conjunction with its use as a skirting for lamps, make this lead one of the most versatile and indispensable of all lead cames.

$\frac{3}{16}$ in.

This, because of its wider leaf, gives more support to the glass and can, therefore, be used not only as an outside lead, but also as an inside one. As a border lead it performs the same function as its $\frac{1}{16}$ in. U-frame counterpart, providing a neat rim to lampshades while, in addition, because of its more substantial channel width, being ideal as a perimeter lead for panels and small windows.

As an inside lead it can substitute for H-frame leads in lampshades or terraniums that are made up of angled glass panels. Where the angle of the joint between the panels is so acute that the glass pops out of the channels in an H-frame lead, then this problem is overcome by bordering the edges of each piece of glass with $\frac{3}{16}$ in. U-frame lead cames, and then soldering them together along their entire length. In this case you are simply replacing the two channels of an H-frame lead with the channels of two U-frames positioned back-to-back, a procedure which allow you to design three-dimensional objects of any shape, however sharply angled. In addition, where two pieces of glass meet at right angles to each other they can be bordered with $\frac{3}{16}$ in. U-frames rather than with H-frames (see chapter 9). Depending on which method you employ, both the empty channels of the H-frame and the edge of the U-lead will be covered with solder when you solder them together, so the end-result will be the same. U-frames are, however, potentially neater to work with in such circumstances.

Two U-frames, when placed back-to-back, can, in fact, substitute for an H-frame lead came in any instance. It is important however not to develop the habit of using them within a lamp, when in most cases an H-frame will house the glass perfectly. Not only will two U-frames give a bulkier appearance to each joint (having the equivalent of two hearts to the one in an H-frame lead), but you will be using nearly double the amount of lead, which means that not only will the finished object be heavier, but, because lead is sold by weight, so it will be that much more expensive.

The $\frac{3}{16}$ in. U-frame is less malleable than the $\frac{1}{16}$ in. and cannot, therefore, negotiate bends or angles as easily. It is, in addition, a difficult lead to cut because it does not have a central strut to balance the pressure of the lead knife during the cutting operation. The leaves tend to crush and collapse sideways, so cut it with a very gentle rocking motion of the lead knife and, when necessary, then restore the channels to their original shape with a lathekin.

Summary

U-frame leads are absolutely essential for any three-dimensional glass projects, and which, because of their non-availability in England till very recently have largely prevented the art of glass from extending beyond the professional studio into the realm of the home craftsmen. They are, however, best used in conjunction with and not in place of H-frame lead cames.

U-frames are also available commercially in America in sizes of $\frac{1}{4}$ in. and $\frac{3}{8}$ in.

There are four procedures that lead came undergoes before it is finally incorporated into a glass project: it must be bought; stored; stretched; and maintained and cut.

1 *Buying lead came*
Plan a project in advance so that you know what size of face width and approximately how many strips (lead came is supplied in 5 ft lengths) are required. Lead is sold by weight, so ordering the wrong pieces can be an expensive business.

For example: a $\frac{3}{8}$ in. H-frame lead, on the one hand, will allow the glass panels in a lamp to be angled much more acutely than a $\frac{1}{4}$ in. lead, but, on the other hand, it would be much too heavy visually for an umbrella-shaped lamp that has less sharp angles. There is, therefore, depending on the shape of the object to be made, a preferred lead width, both functionally and decoratively. The $\frac{1}{8}$ in. difference between a $\frac{1}{4}$ in. and a $\frac{3}{8}$ in. lead may seem infinitessimal, but it is *not*, so try to determine before buying, both what face width of lead came is best suited to the job at hand, and the number of strips needed.

Trundling back and forth to either replace a wrongly chosen size of lead or to buy more of a correctly chosen one (invariably and maddeningly when a project is nearly completed) shows a lack of planning.

2 *Storage*
Try, whenever possible, to keep the lead strips horizontal and straight. Ideally, they should be stored in 5 ft boxes (a box per size of lead came) that are placed either on wall-racks, one above the other, or – to conserve space – under the work bench.

The more you allow the leads to become enmeshed and bent, the more attention you will have to pay to them later with a lathekin.

3 *Stretching*
Always stretch lead came immediately prior to use, preferably with a lead vice (the method is described on page 44). All lead sizes are improved by stretching, although thinner leads elongate more than thicker ones. The process of lengthening both removes all the kinks and convolutions and also hardens the lead itself: unstretched lead lacks the tautness that it requires to be able to follow closely the contours of the glass, being much more hunched and, therefore, unruly. One does, finally, get more mileage, both financially and usage-wise, out of stretched lead as there is, quite simply, more of it.

4 *Lead maintenance and cutting*
It is a good idea to develop the habit of running a lathekin (for the method see page 45) along the channels, ie, *both* channels, of the lead after it has been stretched, whether it needs it or not.

Such an 'open-heart' operation minimises the likelihood of the pieces of glass catching on the leaves of the lead came, which then require additional surgery to pry them open.

After – or (depending on the situation) before – the channels have been widened, then the lead must be cut to fit into its predetermined position in the glass project. Cutting (described on page 43) involves a downwards, oscillatory movement of the lead knife to separate the lead came into two parts with minimum damage to the leaves. If you chop at the lead it will concertina, and this then involves major reparative work to prize open the crushed leaves and channel.

Soldering iron

If there has been any one single vehicle of change that has enabled the art of stained glass to extend beyond the professional studio and into the home, it is unquestionably the coming of age of the electric soldering iron. The traditional method of soldering employed in the stained glass industry by a leaded light maker (also known as a *leaded light glazier*, though this terminology has gradually fallen into disuse) is by the use of a gas soldering iron. This is the time-honoured method; and one that has, for some very good reasons, been resistant to change within the glass industry as such. It is not, though, suitable for a home craftsman. Here an electric one is essential.

The choice of a soldering iron for leaded glass work depends on the following factors:

1 The nature of the work to be soldered – whether flat or curved.

2 The variety of jobs to be covered by one iron – the soldering of lead came joints or the beading of copper foil, or both.

3 Convenience of use – the iron's weight, size and balance.

4 The size of the joints to be soldered – which determines the size of the diameter and the shape of the iron's bit.

5 The rate of working – whether continuous or intermittent, which determines the iron's power input or wattage.

Before describing why, for our purposes, electric irons are best qualified to meet the above requirements it is important to mention various other pieces of soldering equipment that may well be about the house, and which therefore, one may be tempted to use so as to save the financial outlay of an electric iron.

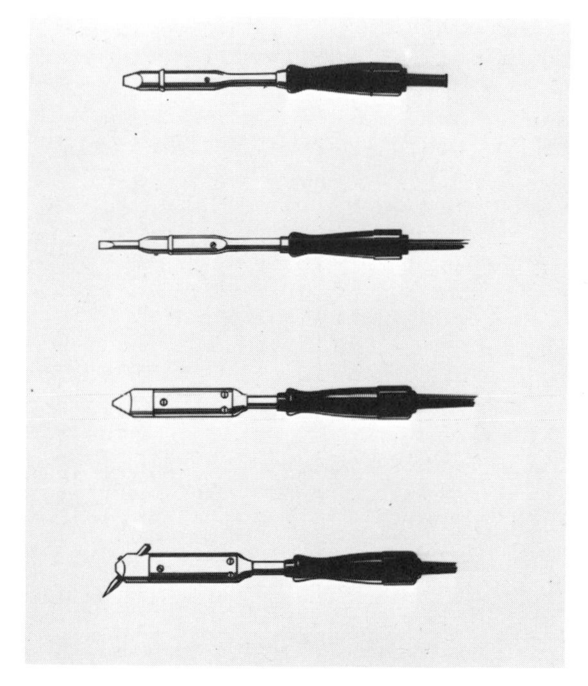

A selection of 'Solon' electric irons suitable for leaded glass. From the bottom, 125 watt pencil bit iron, 125 watt tapered bit, 65 watt straight pencil bit, and 65 watt tapered bit. 'Solon' irons are nickel-plated, but have no inbuilt temperature control
Reproduced by courtesy of GEC-Henley, Gravesend, Kent

Interchangeable parts of a 'Solon' iron: the handle, the bit-holder, the bit, and the heating element. The iron's life is potentially limitless

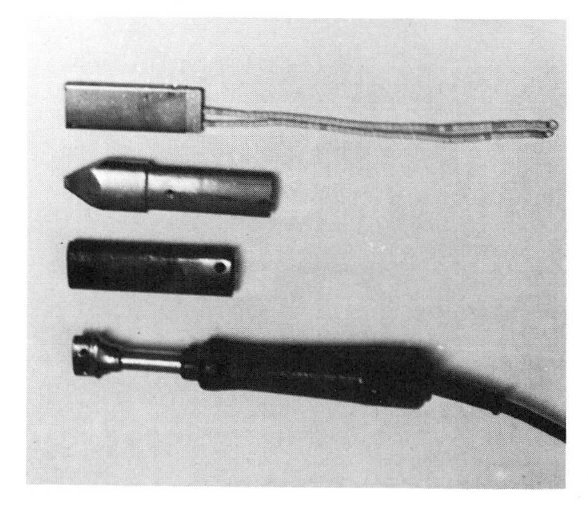

Hand iron

This is a forerunner to the electric soldering iron, but is still used by some people for general soldering work. It consists of a wooden handle attached to an iron shaft on the end of which is the tapered bit which is heated in a stove or oven. When the tip of the shaft is judged to be sufficiently hot then the iron is transferred to the area to be soldered. It does not have any temperature control, is often unwieldy, and also time-consuming.

A hand-iron is, therefore, totally inadequate for our needs.

Blow lamp

This is for soldering large pipes and other metal fittings. It is best left in the hands of plumbers, for whom it is principally designed. If used for soldering cames it would make a searing attack on the joint to be soldered, quite literally stripping the lead came away and cracking the glass: a classic example of burning down the house to roast the pig.

Soldering gun

The gun's primary use is for general purpose spot-soldering work around the house by the handyman and does not, as such, constitute part of our soldering arsenal.

The soldering bits are too small for lead came or copper foil work, while the body is too short and bulky to enable one to probe effectively inside three-dimensional glass projects with any degree of success.

Gas soldering iron

This is not suitable for our purposes. Whereas a gas iron is efficient on flat surfaces, it cannot, however, perform the more intricate soldering required for the inside of curved surfaces. The copper bit on a gas iron is fixed at right angles to the handle and cannot, therefore, operate easily in a restricted area where an electric iron (the bit of which is a straight continuation of the shaft) is much more capable of probing into hard-to-get-at corners.

Another limiting factor is that the flame on a gas iron heats the copper bit which is, as mentioned above, at right angles to the handle and, therefore, to the flame. So it is impossible to use it within the tight confines of, for example, the inside of a small lampshade, because the flame will in all probability crack any glass that is situated at right angles to the joint being soldered. The flame could of course be turned right down when working in such a situation (thereby losing control of the heat) or even a shield could be attached to the iron to prevent the flame from playing on the glass, but the former detracts from the iron's performance and the latter makes it even more unwieldy and bulky.

To sum up: the gas iron is ideal for soldering flat surfaces, but it lacks the qualities needed for soldering inside curved objects. This, when taken with the concomitant difficulties of installing a gas jet in your house, and the inherent dangers of gas as a form of energy, make it impractical for our purposes.

Electric irons

The electric soldering iron has opened the door to the home craftsman. The range and quality of electric irons, interchangeable parts, and ancillary equipment, has increased immeasurably over the years so that the selection now available is bewilderingly wide.

The nature of the soldering operation in leaded glass work requires an electric soldering iron with a heat capacity of somewhere between 80 and 125 watts. The final choice of iron depends, of course, on personal preference, but irons of either lesser or greater wattage do not generally meet these needs.

Irons of less than 80 watts do not usually generate sufficient heat to solder an area as big as a lead came joint though 65 watt irons are usable for pinpoint soldering. They are designed for precision spot-soldering – such as that in radio and television electrical work – and have neither the heat capacity nor a large enough bit suitable for soldering cames. To obtain a soldered joint that is both structurally strong and neat, we need an iron that is, on the one hand, hot enough to melt the solder and, on the other, of sufficient bit size to cover the entire surface of the joint so that it can be bonded in a single operation. The bit sizes of smaller irons prevent them from meeting this last requirement. Whereas they will, no doubt, get the job done in the long run, the finished joint will be a messy, hodge-podge affair due to the piecemeal process of applying the solder 'bit-by-bit'. In the same way that you would not use an artist's brush to paint an area as large as the wall of your house, so avoid lower wattage soldering irons. Fit the tool to the job at hand; not the reverse.

Irons of more than 125 watts, however, can create the opposite set of problems. They are normally supplied with large oval, tapered bits which, due to the bit's greater capacity, generate far too much heat for leaded glass, resulting in the disintegration of both solder and the lead came and, in all likelihood, the work bench underneath. In any protracted soldering session they will, also, for good measure, break the glass and, due to their excessive weight, probably your wrist as well. They are designed specifically for heavy industrial duty work, and are best left to that sphere of operation.

So select an iron in the 80 to 125 watt range to meet your own preferred combination of weight and shape of handle and soldering bit. You will not, of course, be able to make a comparative judgement until you have actually experimented with various models, but experience has shown that women tend to prefer the more manageable, lighter irons, while men often go for the heavier, more substantial 120–125 watt one.

All parts of irons are, incidentally, replaceable, spares being readily available. The heating elements inside the handle, the handle itself, the bit, the bit holder, and the cord, can be purchased over the counter at a hardware store or ordered directly from the manufacturer. This gives the iron a definite permanency, so choose it carefully.

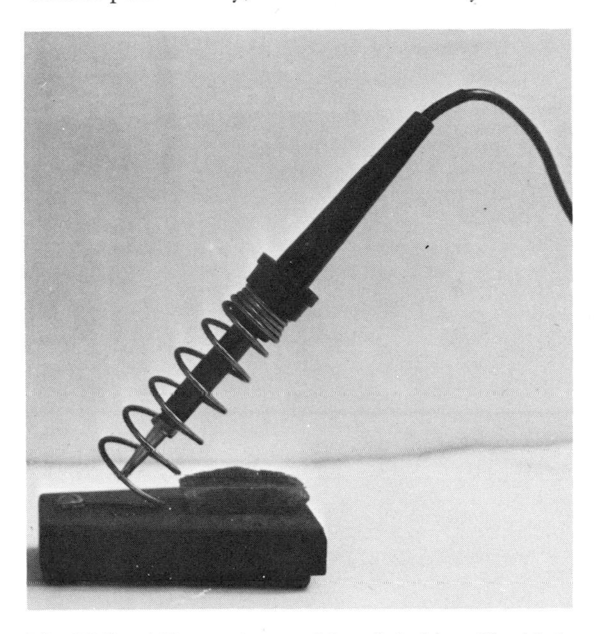

The Weller 100-watt iron and bench-holder. The bit is iron-plated and contains a temperature control unit called a 'sensor'. In America Weller manufactures a similar 80-watt model
Reproduced by courtesy of Weller Electric Ltd, Horsham, Sussex

Choice of soldering iron bit

As a carrier of heat the bit should have a large heat storage capacity, which is a product of its weight and specific heat. The weight is limited by the ability of the operator to wield it and by the need to be small enough to reach the joints in confined spaces, while its specific heat depends on the metal from which it is made.

The above aspects of bits can be described in terms of the shape and metal content of various models.

(a) *The shape*
The standard bit is oval and tapered, but others are pencil- and hatchet-shaped. The choice is yours. Begin, though, with the tapered bit as this is most widespread and has, presumably, therefore proved to be most effective for most users.

All bits, whatever their shape, are interchangeable. They are held in position by a set screw and can, therefore, be removed and replaced – either to insert a differently shaped bit or to replace a badly worn one. This is both an easy and an economic operation, and one which allows you to vary the size and shape of bit to suit the type of soldering that you are doing, such as different angles and widths of lead came.

(b) *The metal*
Bits are made of various metals, the standard ones being:

Copper Copper has, despite its disadvantages in maintenance (see 'tinning' below), one great advantage over other metals: its thermal conductivity, ie, capacity to deliver heat to the job, is much greater than, say, aluminium or iron.

The problem with copper bits, however, is that they corrode (albeit very gradually) during use. One of the prerequisites of the soldering operation is that the molten solder must wet and cling to the bit. Copper is readily wetted, but dissolves in the molten solder, requiring attention in the form of 'tinning' to repair the surface of the bit. (It is the tin in the solder alloy that is the active ingredient that corrodes the copper, hence the derivation 'tinning' to describe the process of restoring the surface of the bit to its original form; of, in actual fact, '*de*tinning' it.)

Maintenance of the bit
The copper bit *must* be tinned before use. This involves first heating it, then fluxing the surfaces, and then running solder over them. The solder should adhere readily to the surfaces, but, if it does not, then the iron is probably too hot (it could, alternatively, be dirty) and must be cooled until the solder sticks. The operation is not complicated, but if you experience difficulty in determining whether you are doing it correctly, then consult your hardware store.

Copper bits do, as mentioned, gradually corrode and oxidise during use, until such time that contact with the work is impaired. It is then necessary to file the face of the bit back into its original shape to remove both the pitted parts and the caked oxidation.

The process is as follows: file down the faces of the bit when it is cold, making certain that they retain their original shape. When they are again bright and even, with all pitting removed, then tin (ie heat, flux and solder) the bit as you did prior to first using it.

As the process of soldering wears down the copper, so retinning will again be necessary. If you use a copper bit then develop the habit of checking its condition before each work session and tinning it whenever corrosion shows. This ensures its continual good quality, and prevents the cavernous pitting that occurs from total neglect, and which is invariably irreparable when you do, finally, get round to fixing it.

Nickel and iron-plated All soldering bits are made of copper, but some have a thin layer of nickel or iron that acts as a protective coating to prevent the dissolution of the copper by the tin in the lead.

A nickel-plated bit is longer-lasting than the purely copper one and does not require to be tinned (reshaped) at all. In fact, filing it down would destroy the protective nickel coating and ruin it. To maintain it, all that is required is that it is brushed lightly with a wet cloth or sponge when necessary to remove any excess solder or dirt that may accumulate on the bit, and which would thus impede the soldering operation.

The iron-plated bit is the most durable of all, and is, in addition, supplied by one manufacturer, Weller, with an in-built sensor in its bit. The sensor consists of a magnet that is attracted to a temperature sensing element when the bit is cold. This pulls on a power switch, and as the bit reaches the marked temperature (the Curie point) the temperature sensing element can no longer hold the magnet, which retracts, thereby pushing the switch to the 'off' position. When the tip cools slightly the temperature sensing element again attracts the magnet to resume heating.

The Weller iron-plated iron is, therefore, manufactured with a heat-regulator which, although not as accurate as a variable resistor (see below), does restrict the temperature to certain predetermined limits. One of the Weller irons is a 100-watt one that has a bit that is heated at an average temperature of 315°C (599°F). This is below the melting point of lead (327°C – 620°F), and so it is impossible to melt the lead cames when you are soldering them: which, as any veteran glass exponent will appreciate, eliminates a vast amount of the problems that can occur in glazing. This fact, in conjunction with the durability of the iron bit, makes the performance of the heat-regulated iron the best of all the electric soldering irons for our purposes: quite formidable, in fact.

The iron-plated bit is maintained in the same way as the nickel one: just wipe it with a wet cloth. Do not file it.

Temperature control

A complaint frequently lodged against electric soldering irons by glaziers in England is that they are underpowered; that they cannot maintain the heat necessary to solder a considerable number – sometimes literally hundreds – of joints uninterruptedly. I personally have seldom experienced this problem (ie, of their losing heat: the problem for me is more frequently of their becoming too hot) with electric irons, and it is improbable that you, unless you are involved in large commercial projects, will encounter the same handicap. There are, however, three methods of overcoming temperature variation in the iron.

1 The most unsophisticated method is the plug-and-unplug routine. When the behaviour of the solder indicates that the iron is too hot, then simply unplug it and let it cool for a while before replugging it. This is, at best, a 'guesstimate' procedure, but perfectly legitimate. It is, however, very advisable, on replugging the iron, to test it with the stick of solder away from the work in progress before returning to do the joint you were previously working on. This routine can, however, prove tedious in the extreme. If the wall switch is not situated immediately within arm's reach, then it will be necessary to stretch or bend to locate it, probably round or under various table legs and other oddments.

2 In order to eliminate the above inconvenience, attach an 'on-off' switch to the cord on the iron. Pay attention, though, to where it is positioned: if it is too far from the iron you will have defeated the purpose for which it was installed, yet if it is too close to the iron it can get entangled in the actual work or in the tools and materials being used. Determine, before positioning it, where it is most likely to be accident-free.

3 The use of a variable resistor (known as a *rheostat* in the United States). This really is the cure-all for all difficulties in soldering caused by temperature fluctuation. The variable resistor controls the amount of heat being fed into the iron and, of course, does away with both the plug-and-unplug and cord switch routines. It is a small instrument that, like the switch, must be positioned on the cord at a suitable distance from the iron itself. It contains a dial which regulates the temperature so that it remains constant at the desired level.

The variable resistor is the electric soldering iron's ultimate answer to all criticisms levelled against its inability to provide the correct amount of heat: not only will it keep the heat constant, but it will, by a twist of the dial, decrease or increase it to an exact (unlike the air-intake sleeve on a gas iron, where any change in temperature is estimated), predetermined level.

Professional glaziers have to handle a large amount of soldering. A great deal of this entails soldering several different sizes of lead came within a single window or panel. For example, the perimeter lead came might be $\frac{1}{2}$ in., while there may be two or three leads of different width, such as $\frac{1}{4}$ in. and even $\frac{1}{8}$ in., within the window itself. The heat required to solder a joint depends on the surface area of the metal to be bonded, and so one cannot apply the same heat to these various widths of lead. On the one hand the increased temperature required to solder the $\frac{1}{2}$ in. lead would, if used on the $\frac{1}{8}$ in. joint, melt the lead came, while, on the other, the correct heat needed to solder the $\frac{1}{8}$ in. joint would be insufficient to solder the $\frac{1}{2}$ in. one, so obviously, a rapid and controllable change of temperature is a prerequisite in such situations.

'Strike while the iron is hot' is a good maxim, but the iron must not be allowed to become too much so. Likewise, an iron must be able to keep its cool (and, thereby yours, incidentally), yet here again, it must not be allowed to become too cool. The variable resistor provides the balance between the two, and is a prerequisite for anybody doing a large volume of soldering work: one which, when combined with an iron-plated iron, is the quintessence of problem-free soldering.

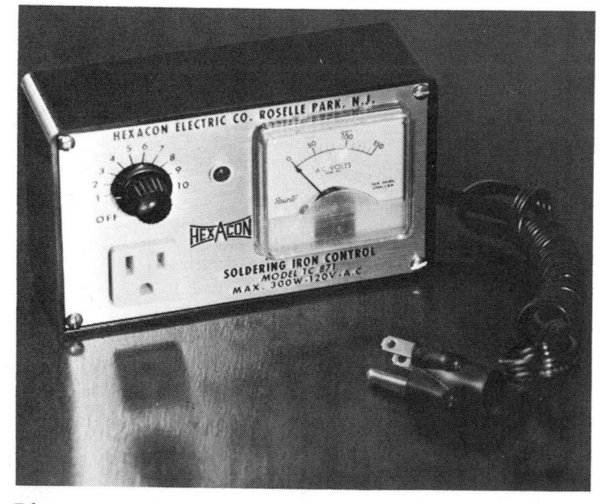

Rheostat temperature control unit manufactured in America by Hexacon

Bench-holders

These are available in the form of a funnelled spring for smaller irons (up to, and including, 80 watt) or as a stand for larger ones (100 watt and above). They are not always available commercially, and if you cannot purchase one at a hardware shop then either order directly from a manufacturer (see Suppliers list), or improvise by making a cradle or bracket out of bent wire.

Consider the bench-holder as an extension of the iron itself; as an integral part of it, in fact. In this way you will develop the habit of always placing the iron on it between use, and not on the nearest glazing tool, piece of glass, or window sill. Irons have a burning desire to make their mark on things, and there is very little that an iron touches that will rise phoenix-like from the ashes for a second life. So use a bench-holder.

Choice of soldering iron

All irons, when correctly maintained, will perform satisfactorily, but some, as explained above, have a potentially shorter life-span and require more attention than others – there being, not unexpectedly, a correlation between the longevity and the respective price of each iron. Try to push your budgetary constraints to their limit in choosing an iron. One that performs well is, along with the choice of glass cutter, of paramount importance, and its effectiveness will be self-evident in the finished work. I do not feel that it is necessary to progress from a copper iron to an iron-plated one 'bit-by-bit', so to speak, in order to master the art of soldering. Such 'through the mill' apprenticeship is of no value: on the contrary, as plated irons require the least maintenance, so they are the easiest to use, and should, theoretically, be the ones with which to start.

The final selection does in part depend on the quantity of soldering that you will be doing: obviously, if the output is minimal, then a copper bit will suffice. When, however, your work involves a considerable amount of soldering, then a plated iron is preferable, if only to avoid the time-consuming tinning of copper bits. All irons, whatever their bit, will, if correctly maintained, give long service, but with a substantial amount of soldering the stage will soon be reached where the benefits gained from the money saved from buying a cheap iron will be outweighed by the valuable time lost on continually tinning it. My own recommendation for the newcomer (until, at least, he or she has had time to experiment with different makes to determine his preference) is that he use the Weller 100-watt iron regulated at 315°C (599°F). This, as mentioned, provides enough heat to readily melt the solder but not enough to do so to the lead cames, and it does, as such, eliminate probably the most disheartening aspect of soldering that a beginner can experience; namely, to burn away the joint that he is trying to solder. The 100-watt iron is, in addition, made to take four different bit widths, from $\frac{1}{8}$ in. to $\frac{3}{8}$ in.; the correct interchange of which for the job at hand will cover virtually all of an amateur's soldering requirements. Do not, finally, buy a variable resistor until such time that the volume and complexity of your work demands many hours of continuous soldering and frequent temperature changes. It is mentioned here more for the professional glazier.

Solder

Solders are metallic substances of lower melting point than the metals that they join, and act by:

(a) flowing between the metal surfaces which remain unmelted;
(b) completely filling in the space between the surfaces;
(c) adhering to the surfaces;
(d) solidifying.

The composition of the solder most suitable, ie, that is commercially available, for leaded glass work is an alloy of 60% tin and 40% lead. Its behaviour as regards melting and solidifying is most easily shown by the diagram on page 76.

The entire range of tin/lead solders is depicted with pure tin on the extreme right and pure lead on the extreme left, while temperature is represented in Centigrade and Fahrenheit on either side of the diagram by the height above the base line. Pure tin has a melting point of 232°C (449°F) and pure lead one of 327°C (620°F). Mixtures in all proportions have their melting points indicated by the curved lines AE and EB.

Take, for example. a combination of 20/80 (20% tin, 80% lead). Such a combination would have to be heated to 280°C (536°F) for it to be completely liquid. This temperature is almost as high as the melting point of 100% lead, and as lead came is almost pure lead, so, in order to melt 20/80 solder, the likelihood of melting the lead came itself would be great. In addition, a 20/80 solder takes from 183°C to 280°C (361°F to 356°F) to become completely liquid, during which time it is in a pasty form. When the combination reaches melting point, 280°C (536°F), it then has to pass back through the pasty range before solidifying (solders are devoid of strength in the pasty range as they are not truly solid). The whole process would be far too time-consuming and cumbersome for the rapid, precise soldering required in leaded glass work.

Point E, however, is of cardinal interest. It is evidently the easiest, ie, lowest, melting point of all combinations of tin/lead solder, and for this reason is known as the *eutectic*. It contains 63% tin and 37% lead, and melts at 183°C (361°F).[1] A peculiarity of the eutectic mixture is that it is the only composition which transforms from the liquid state (*liquidus*) to the solid state (*solidus*) without passing

[1] The latest British Tin Research Institute figures indicate that these may be nearer to 62% tin and 38% lead.

through an intermediate pasty form and behaves similarly, therefore, to a pure metal. At 182°C (359°F) eutectic solder is completely solid, at 184°C (363°F) it is entirely liquid, while at 183°C (361°F) it may be either liquid or solid or a mixture of both in any proportion. This means that because the solid-liquid-solid transition is practically instantaneous, so soldering occurs with maximum speed and thereby minimizes the risk of movement in the parent grafts (lead came or copper foil) during the soldering operation.

Solderability is closely connected with wettability, and the eutectic solder wets the metal that it is joining at a lower temperature than any other combination of solder. Solders with either more or less tin than the eutectic, or containing antimony or impurities, are less easy to work with.

Solder is available commercially (in both England and America) in a combination of 60/40. This means that there is a working zone of 5°C (9°F) in which the solder is in a partially pasty state, but as this discrepancy in temperature is, in fact, negligible, so 60/40 is ideal for our purposes.

A 50/50 solder can be purchased in the United States. This gives a working range of 30°C (54°F) which some leaded glass practitioners prefer for copper foil work only (ie, not for lead came soldering) as it enables the solder to flow along the copper foil for a few extra seconds before completely solidifying. This helps one, they maintain, to obtain a well-rounded, uniform ridge of solder along the joints. A 60/40 solder does, though, serve adequately.

Solder for leaded glass work can be bought from ironmongers in England in either stick or bar form. The former is preferable as bar solder is too thick for pin-point soldering and results, on being melted, in too much solder being deposited on the joint. This, in turn, overlaps the glass itself – causing a loss of definition – or leaves a large, conspicuous blob of solder on the joint, which, when you try to coax it into a less evident shape with the soldering iron, invariably responds in a most unco-operative manner. So use the stick solder (known as *blow-pipe solder* in hardware stores) if you can get it. It is supplied in sticks about $\frac{3}{16}$ in. wide.

Solder in America is available in $\frac{1}{8}$ in. coils. Solder is, incidentally, because of its high tin content, not cheap. Develop the habit of saving all small pieces and then remelting them later.

Core solder

These are hollow or grooved wires or tubes filled with flux. The flux is introduced warm as a paste during the extrusion process of manufacturing cored solder, and, when cold, solidifies within the solder. Cored solder is not for this type of work; however much of a time-saver it may seem. It is made principally for electrical work, and leaves a resinous gum on the surface of the lead and glass that is only completely removable with a solvent such as methylated spirits. It does, in addition, build up a layer of resin on the tip of the soldering iron – impairing its performance – while, finally, if further corroboration is required to dissuade you from using it, generating clouds of noxious smoke.

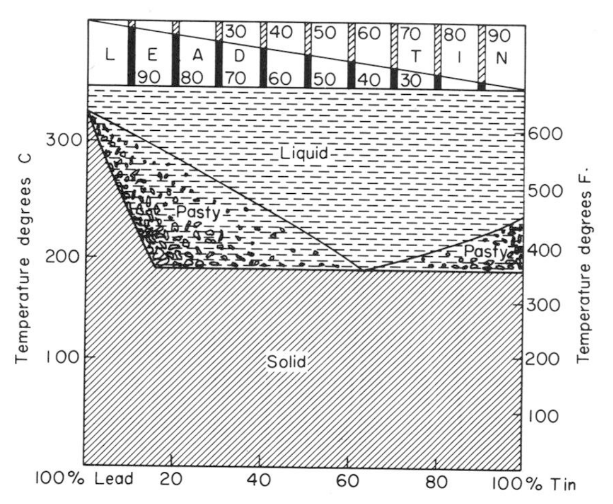

The melting range of tin/lead alloys
Reproduced by courtesy of the Tin Research Institute

Flux

This is a substance that is applied to lead cames after they have been cleaned, which both prevents them from oxidising while being soldered and promotes wetting by the solder. Lead came oxidises when exposed to the air – a thin film of oxide forms on the surface of the lead – and solder will not adhere to the lead until this film has been removed.

It seems logical that, having cleaned the lead surface, one could immediately solder before the oxide reforms, but the heat generated in soldering makes the lead even more reactive than usual to oxygen, thus accelerating the process and sabotaging any such procedure. So first the lead must be cleaned, and then the flux is applied to seal off the surface from the air. Flux acts, in a way, as a catalyst between the solder and the lead.

A flux should have the following characteristics:

(a) It should form a liquid cover over the lead came so as to exclude the air.

(b) It should continue to do so up to soldering temperature.

(c) It should dissolve any oxide on the lead came or solder, and carry these unwanted materials away.

(d) It should be displaced from the lead surface by the liquid.

The illustration on this page shows what actually transpires during the fluxing process. All fluxes perform the same basic function, but they vary in character and strength to meet the specific soldering circumstances at hand. Certain metals require a very strong flux to bond them – for example, hydrochloric acid in water is a suitable flux for soldering zinc – but it leaves a residue that is strongly corrosive. This would be most unsatisfactory for our purposes as the flux would eat into, and eventually pit, the lead came.

Ideally what is required is a non-corrosive flux, but this is in the nature of an anomaly as fluxing action demands some degree of reactivity between flux and metal. There are substances, however, which are sufficiently active at soldering temperature to act as fluxes. Fluxes tend to fall into three categories: corrosive, intermediate, and non-corrosive. The aim is to use the mildest flux possible that will do the job adequately, and tallow and oleic acid (which qualify as intermediate fluxes) meet the requirements for lead came very satisfactorily. Your ironmonger will always be able to advise you if you experience difficulties in knowing which flux to buy: just make certain that he understands that you are using 60/40 solder to join lead came.

Fluxes are available in either liquid or paste form, and both are suitable, though I would tend to recommend the paste. This is because liquid fluxes are liable to drain off the surface of the joint, leaving insufficient flux behind, and to run over the glass and through on to the work surface underneath, which, if it is your work drawing, will tend in the case of oleic acid to erode and ultimately disintegrate.

Whichever is used, whether paste or liquid, should be applied with a hard bristle brush (see page 78) so that the flux can be scrubbed right into the newly cleaned joints. Old aerosol containers are sometimes refilled with liquid flux and used to spray it on to the joints. This makes for rapid work, but does not ensure that the lead is as thoroughly fluxed as when it is brushed on. In addition, spraying is, by its very nature, an indiscriminate operation that results in not only the glass project being fluxed, but also the entire immediate environment.

If using liquid solder it is best kept in a jar in very small quantities as then it will not, if inadvertently overturned, swamp the entire work area.

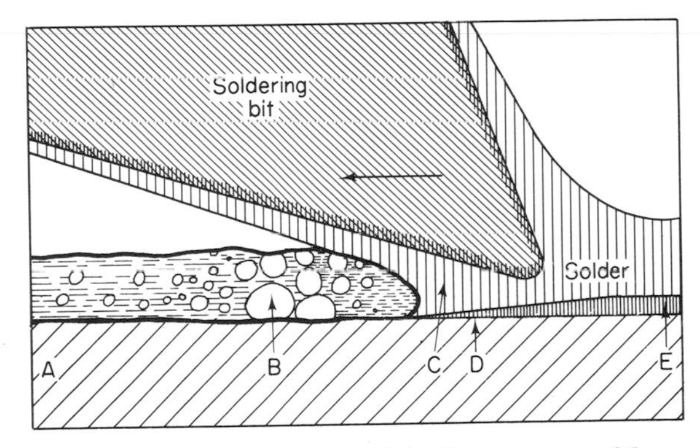

Diagrammatic representation of the displacement of flux by molten solder

A De-oxidized metal in contact with the flux
B The flux
C Liquid solder displacing the flux
D The tin in the solder reacting with the lead came to form a compound
E Solder solidifying

The soldering process

Lead came soldering

It is the essential feature of a soldered joint that each of the joined surfaces is wetted by a film of solder and that these two films are continuous with the solder filling the space between them.

It is not simply a matter of applying the solder so that it bridges the gap between the surfaces so that they adhere together. It will soon be discovered that solder has a distinct character of its own, with idiosyncrasies that will generate feelings of ambivalence on the part of the craftsman. When it co-operates to form rapid, flawless joints you will love it, but there may be times when it decides to act in a contrariwise manner; being by degrees mischievous, capricious, and sometimes even verging on the satanic. You may well wonder, as you battle to smooth a militant bit of solder, which one of the Fates you have unintentionally angered, and why.

Although the fault may appear to be manifestly that of the solder – in that it is the solder that is seen to malfunction in some way – it may be doing so for any one of several reasons for which it is not itself to blame. There are four components that combine to form the soldering operation – lead came, the soldering iron, flux, and solder – and if one of the other three variables is not performing its specific function correctly, then this will impair the ability of the solder to bond itself on to the joint. The step-by-step process of soldering is now described so as to minimize the eventuality of any potential mishaps.

(i) *The lead cames must be shaped so that they fit together*

The endings of the strips of lead came should be cut so as to facilitate the soldering process; the parts to be joined should fit closely so that the space between them is narrow enough to become completely filled with molten solder (which is drawn in by capillary force as the solder wets the metal), but not so wide that a large amount of solder is required to bridge the gap. Experience will show what is, and what is not, too wide a distance: when it is too wide the solder will flow right through and solidify on the underside of the joint, leaving a gaping hole on the top. And when you come to turn the panel over to solder the other side, then the large lump that had formed there due to an initial attempt to solder the top side will, on being remelted, promptly disappear back through the gap to re-form underneath. And so it goes on. You will find yourself involved in a hide-and-seek exercise as you chase the solder back and forth through the joint.

To avoid this, use a sharp lead knife to cut the lead endings so that they fit snugly together, when necessary mitering them at an angle so that they are flush against each other. If, however, on completion of the leading stage, a large gap does exist, then rather than dissemble the project, cut an extra piece of lead came and fit it into the hole. This will act as a stepping-stone for the solder between the original lead endings.

(ii) *The surfaces to be joined are specifically cleaned*

All the oxide, grease, and dirt, must be removed from the area to be soldered, as these, being unsolderable, act as a barrier to the soldering process. The best method of cleaning the lead is to scrub it with a wire brush (a copper-wired suede shoe cleaner is suitable). Do not apply too much force as this will result in the leaves of the lead becoming torn. The object is to remove the top layer on the lead (ie, the oxide) so as to leave a clean, milled surface which will accept the solder. Too much pressure will result in the leaves being ripped off, and could, in extreme cases, scratch the surface of the glass itself.

If you do not possess a wire brush then use something sharp such as a razor blade, wire wool, or knife. Do not, in any event, use a degreasing solvent. Alkalines, such as a mixture of sodium salts, are used as detergents in some soldering work, but are unsuitable for our purposes as they are difficult to obtain and are often poisonous.

(iii) *Soldering flux is applied*

Solders are unable to wet a metallic surface which, having been exposed to air, has already entered into a chemical combination with oxygen to form an oxide film.

The next stage, therefore, after cleaning, is immediately to apply flux to seal off the deoxidised surface from the air. This operation will, however, be rendered totally ineffective unless the cleaning in (ii) above has been done thoroughly: the melted solder will simply roll off the fluxed joint, no matter how much flux you have rubbed on to it. So first clean the joint, and then brush on the flux.

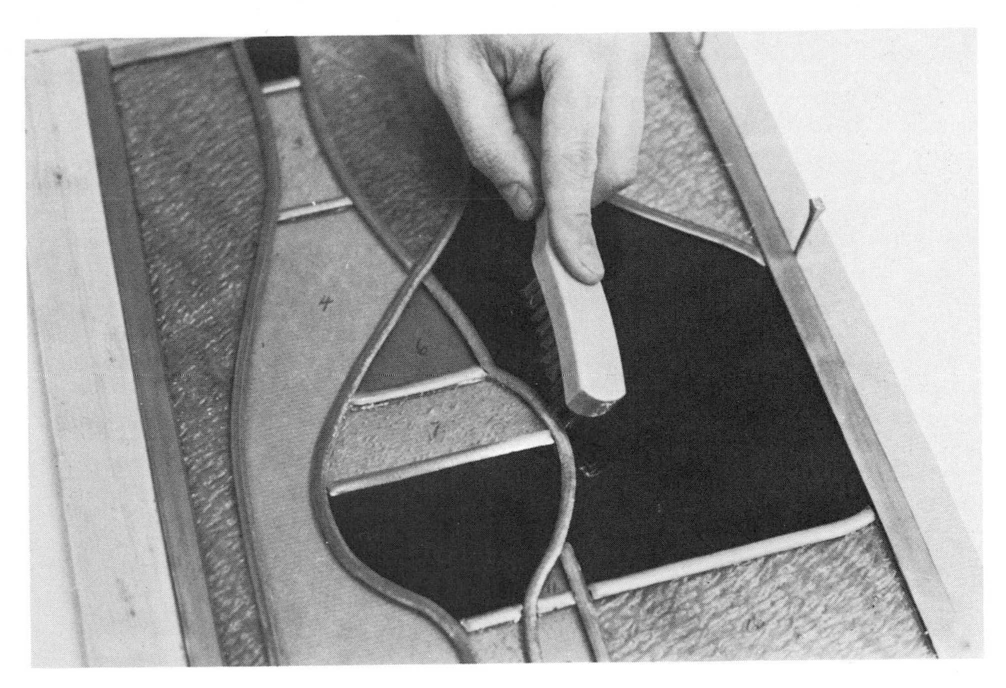

Rubbing the joints with a wire brush to remove the film of oxide

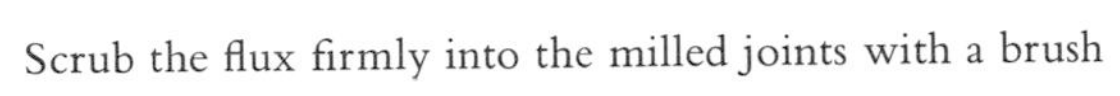

Scrub the flux firmly into the milled joints with a brush

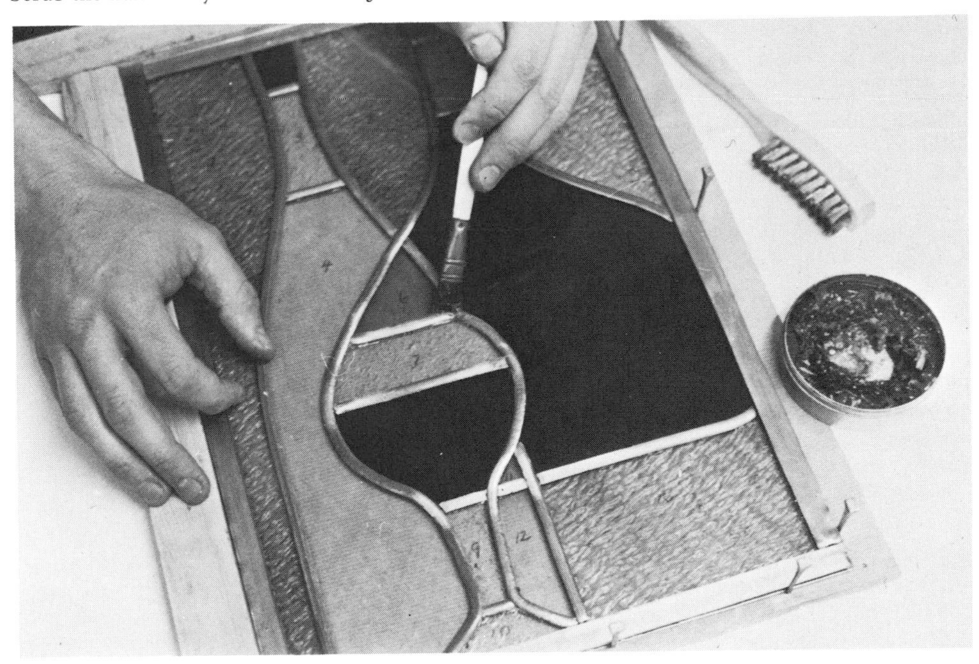

(iv) *Molten solder is applied*
The actual soldering of a joint occurs when the stick of solder is placed on the intended joint and touched with the tip of the soldering iron. The solder should melt and flow smoothly across the joint. Then raise the iron vertically off the solder. The heat from the iron should melt the end of the stick so that a blob of solder transfers, ie, is fed on to, the joint so as to bond it.

Remember that, provided that the joint space is full and that there is sufficient solder to round out, and thus reinforce, any sharp corners, an excess of solder does not add to the strength of the joint. So if you are uncertain as to how much solder to melt off the stick, it is better to apply too little rather than too much: it is easier to keep adding small amounts of solder to a joint than to attempt to try to spread out a large mound of it. In such an event you either have to try to run it back along the length of the strips of lead came that you are soldering (invariably a messy job as you probably did not clean and flux as far back along the lead came as you are trying to steer the solder), or else to sweep the excess solder off the joint on to the glass (to be cleaned off later). Neither remedy tends to be satisfactory as the more you fuss around the joint with the tip of the soldering iron, the more likely it is that the heat generated in such an activity will melt and dissolve the lead cames themselves. The problem will then be compounded in that you will be trying to solder a joint that, in fact, no longer actually exists. So apply solder cautiously and minimally.

If difficulties arise in soldering try to determine systematically where the problem lies. If the joints are neatly mitred, well cleaned and fluxed, and the solder is a combination of 60/40 tin and lead, then the fault lies in the soldering iron: either in its physical condition or in its temperature.

Firstly, its physical condition:

1 Check whether the tip of the iron is clean. Solder will not transfer from the surface of an iron that is dirty because, as the dirt is unsolderable, so it creates a barrier to the soldering process.

2 If the iron is plated then there may be surplus solder on the tip which interferes with the performance of the newly melted solder. Clean the tip by brushing it gently with a damp cloth or sponge.

3 If the bit is copper then it may need to be retinned. Examine the surface of the tip for any signs of corrosion. If it is pitted then this will limit the transfer of heat from the iron due to the reduced surface area that comes in contact with the solder. If this is the case, then file, flux, and, finally, tin it.

Secondly, its temperature:

1 If it is too hot then it will not transfer the solder on to the joint. The solder appears to evaporate or to disintegrate into a multitude of small globules. In addition, the iron will, no doubt, burn away the lead came itself as you apply the solder (the melting point of lead cames is, remember, not so very much greater than that of the tin/lead alloy).

2 If the iron is too cold then it will not pick up and transfer the solder adequately. The solder that does transfer will solidify before you have time to fashion it into a neat shape, giving a peaked, stalagmite appearance to the finished joint.

Before starting to solder make quite certain that you have allowed sufficient time for the iron to heat up. Test it, if necessary, on a scrap piece of lead before commencing on the panel. This will eliminate the possibility of spoiling your first few joints if the iron is not at the proper temperature.

(v) *The joints must be cleaned*
When all the soldering has been completed – on *both* the front and back, or inside and outside of the object that you are making – then any residue of flux, pieces of solder, and general dirt, must be removed.

There is no need to wait for any specific amount of time to allow the soldered joints to cool down. Cooling occurs instantaneously as the solder solidifies by conduction of the heat from the joint into the adjacent cooler lead came. So as soon as the soldering operation is completed you can start to clean whatever you are making.

The use of a scrubbing brush with a detergent (such as washing powder or a general household cleaner) should remove the dirt from the lead and glass. Scrub the detergent into the joints thoroughly and then wash it off. When it is dry make a final examination (again, on both sides of the project) of the lead and glass to make certain that all the grit and flux has, in fact, been dissolved and removed by the cleaning action. The panel is now complete unless it needs (depending on the intended end-use) to be weatherproofed by cementing, to have patina applied, or to be blackened with an antiquing gum.

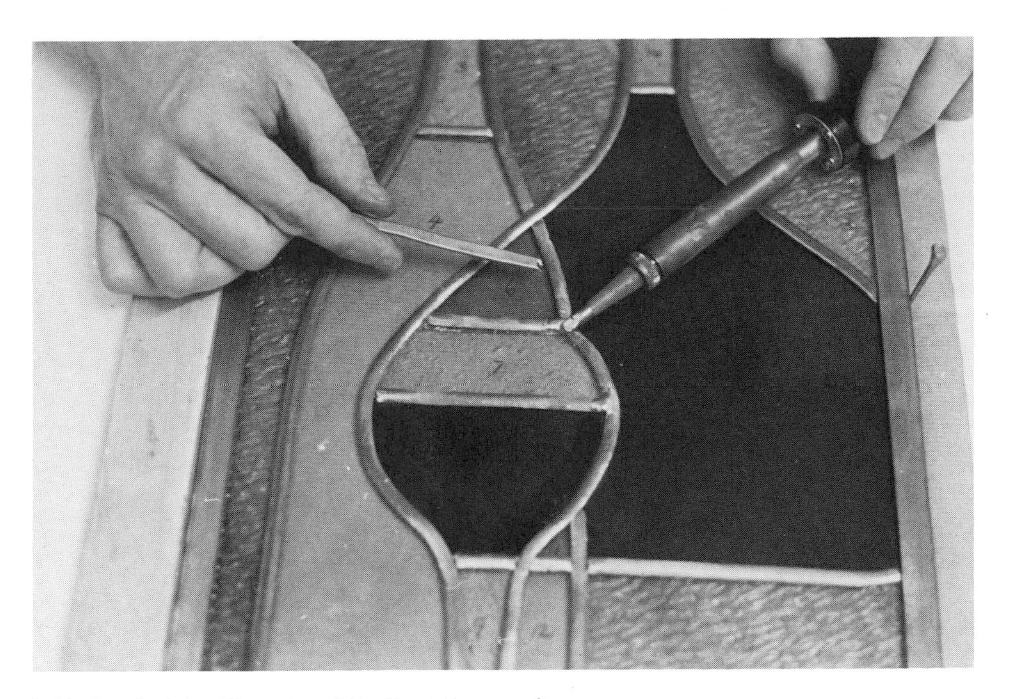

Soldering the joint. Place the solder directly over and on the joint and bring the bit down vertically onto it

Chapter 5

The technique of cutting glass

Probably the most difficult aspect of making leaded glass objects is the time spent on learning how to cut glass. It can be a most frustrating experience in terms of time wasted, cost of glass, and pain, so special attention should be paid to perfecting this process.

Glass as a medium can be most unco-operative; a piece that appears correctly cut will, on occasions, break at the wrong point, rendering useless a specially chosen, ie, matching, piece of glass. Not that it is any consolation, but the fault is not always that of the person cutting the glass.

Glass is tempered during manufacture by a process known as *annealing*. The glass is slowly cooled so that its molecular structure can gradually fit together correctly. When glass is annealed too quickly or too slowly, it retains some molecular tension which is due largely to the different coefficients of expansion of the various oxides that go to make up the particular colours in each sheet. The more colours, ie, the more oxides, then the more potential innate tension that is likely to exist. With experience, a 'feel' for incorrectly annealed glass is developed; it has a certain 'nervousness' and often shatters when scored. In extreme cases there is nothing that can be done about this. However well-scored the surface of the glass is, the break will still go its own way. The likelihood of such intransigence is reduced by scoring the glass correctly, and the following instructions will, I hope, help to minimize such mishaps, though they still can happen.

The correct way in which to hold the cutter

This should be held between the first two fingers – the index and middle fingers – and steadied underneath by the tip of the thumb. This is not a position that is naturally comfortable for everyone at first, but is well-proven to provide the best balance of control and pressure. So persevere. This is not to assert that a cutter will not function if held in any other manner, for example, as you would a pen. It will. Such a position can still, of course, score the glass, but the action of cutting should ensure that the cutter is held so that the wheel is perpendicular to the surface of the glass so that it makes continuous, smooth contact with the glass as it rotates. Held at an angle of less than 90°, the wheel is less easy to guide and more likely to skid along, rather than to cut, the glass.

It is important, initially, to dispel any feelings of apprehension about the dangers of working with glass; which is sharp, awkward to handle, and sometimes capricious. It can be all of these if it is allowed to work *against*, and not *for*, you. It is vital from the start to adhere to the rules. One of the most important of these is the manner in which the cutter is held. The recommended position gives a continuity of movement from the shoulder down to the wrist and takes pressure off the fingers. You will soon know if you are using too much wrist in your cutting: the finger joints become stiff and painful, and the skin between the fingers becomes chafed and inflamed. If this occurs then reassess your position; you are probably holding the cutter at an angle and not vertically, and are hunched up over the work surface, thereby not making use of your shoulder, but taking all the pressure in the wrist and forearm and losing mobility as well as having to shift your body about in order to cut angles.

To start with, get hold of some pieces of glass, preferably offcuts. Clear glass in sizes of 6 in. square and upwards is very suitable. Put one of these on the work surface and hold it securely with one hand while cutting with the other. Make sure before beginning that the surfaces of both the glass and work bench are clean. Any small particle of dust or glass on the former can make the cutter skip or be thrown off its intended course, thereby causing an incorrect break, while any object on the work surface, however miniscule, leads to uneven pressure on the piece of glass being cut, often causing it to shatter. Place the cutter about $\frac{1}{8}$ in. from the edge of the glass furthest from you and draw it towards you. Run it off the nearest edge of the glass remembering not to ease off the pressure on the cutter as you reach the bottom edge. The entire score should be made with equal pressure as any unevenness can result in unpredictable breaks. Now go back to the point on the glass where you began the score, and, starting exactly where you did before, this time run the score line backwards off the far edge, ie, over the remaining $\frac{1}{8}$ in. thus ensuring that the glass is scored for its entire length.

It is possible that your first attempts to score the glass will result in your not actually breaking the surface of the glass at all: you may find that the cutter skids over the slippery glass surface without making a cut, and, that when you try to cut a curve, you end up in contortions. Check your stance, whether you are holding the cutter correctly, and whether the cutter is sharp enough. Some glass, such as flashed glass, is much more difficult to score on one side than on the other. (See the end of this chapter.) This is because the smoother side is harder and has a brighter and more slippery sheen than the other. If you are unable to score it satisfactorily, then turn it over and try on the duller side.

Try to maintain a constant pressure while making the score. Too much pressure produces a 'shower' of glass particles, while too little will not actually penetrate the surface of the glass. A good method of determining whether you are applying pressure constantly is to close your eyes and listen to the 'sing' of the glass as you cut. It is not necessary to apply great strength – on the contrary, this will

Front view of the correct way in which to hold the cutter

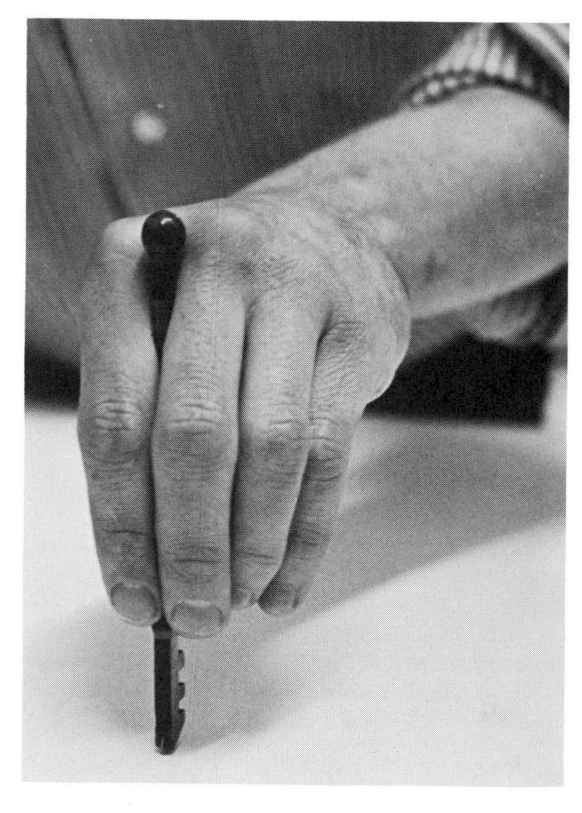

Side view. Notice that the cutter is held vertically

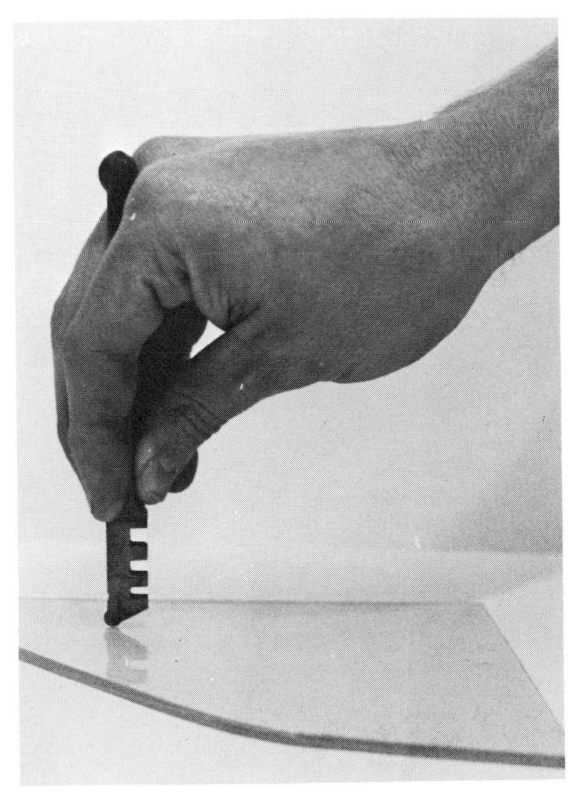

Experimenting on an offcut of glass

cause the cutter to 'dig in' and create unnevenness while blunting the cutter. Concentrate on maintaining a fluid movement so as to give an even, crisp bite, while remembering to keep the cutter in a vertical position. Practise by making as many cuts as possible on the piece of glass. Vary the speed and pressure of the score so as to obtain the optimum effect. You will soon get the feel of what is, and what is not, a clean score.

On *no* account go over, ie, re-cut, a score. This can cause the glass to break incorrectly, and will also blunt the cutter.

By the time that the piece of glass resembles a shattered car windscreen, it is time to discard it and start on a fresh piece. (Remember to brush clean the work surface.) Now make a score down the middle of the glass, again by drawing the cutter towards you. The score should be crisp and even and of equal pressure, and should run right off the edge of the glass nearest to you. If you feel that you have done this correctly, then pick up the piece of glass and, holding it with both hands so that the score is in the middle, clench your fists and make a sharp downwards and outwards movement with both hands so as to snap the glass along the score. It should break cleanly and easily. If not, then try to determine the reason. Check whether you have applied sufficient pressure to break the glass or whether there are skips in the actual score. If necessary, repeat till it comes easily.

Try to determine which balance of speed and pressure gives the smoothest cut

Completion of the initial cut. The score line is visible on the glass

Holding the piece of glass prior to breaking it in two

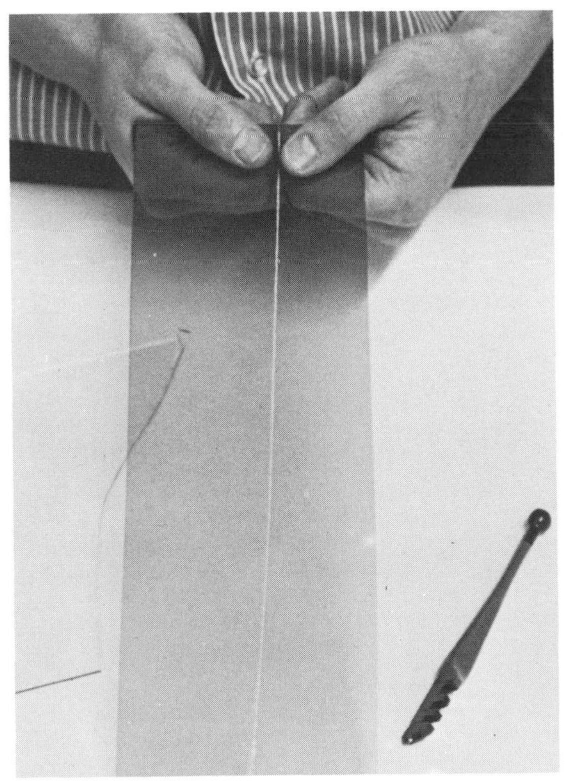

Place your hands on either side of the score line and snap the glass with sharp downwards and outwards movement

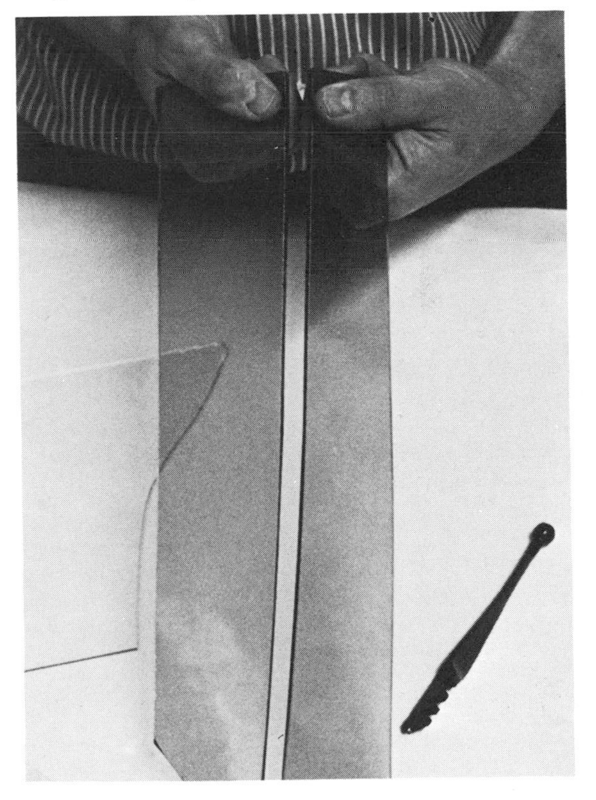

Cutting curves

The next stage is to cut curves. Start with gentle ones and then progress to sharper angles. You will soon discover what can, and cannot, be achieved. This is a good stage to introduce the method of tapping. Make a series of curved scores in ever-deepening segments, and then tap each one and remove it with the glass pliers. Tapping involves using the ball-end of a glass cutter (or any other suitably round instrument) to fracture the glass along the score. Tap on the underside of the glass under the score. It must be *directly* under the score and not to one side as such pressure is totally ineffective and may cause a crack to run off from the score itself. Tapping produces a crack that will 'run' the length of the score, thereby breaking, or rather weakening, the glass. Hold the glass so that both sides of the score are securely clasped in one hand while the other does the tapping. If the glass on only one side of the score is held and the other side is unsupported, then if the tapping causes the glass to snap suddenly the unsupported piece will fall, probably breaking. Not only will a useful piece of glass have been wasted, but one's reflexes often cause one to grab involuntarily at the falling piece so as to prevent just such an end. The agility displayed in intercepting a falling piece in mid-air is not usually compensated for by the likelihood of severe cut fingers obtained in the process. So don't be tempted. Hold the glass correctly while tapping.

Tapping does not, it should be emphasized, entail the use of force: the break in the glass has already been made by the score, and the fracture made by tapping only helps to deepen the crack. Do it gently, otherwise it may shatter. Progress can be checked by holding the glass up to the light while tapping; you will see a shadow form and lengthen in the score as you tap along it.

When all of the curved segments have been tapped, remove them, one by one, with the glass pliers. They should separate (if they have not already done so due to the tapping) with minimum pressure from the nippers. All that should be required is a gentle, horizontal, pulling movement with the nippers to separate each piece. Tapping tends to leave a jagged edge of small, sharp, splinters on the glass, and these should be removed by chewing away with the grozers or by rubbing the edge of another piece of glass along the edge to knock them off.

Tapping should only be used for sizes of glass that can be held in one hand; obviously if the piece is too large then it is not possible to hold on to both sides of the score. In such cases one of the following methods should be used:

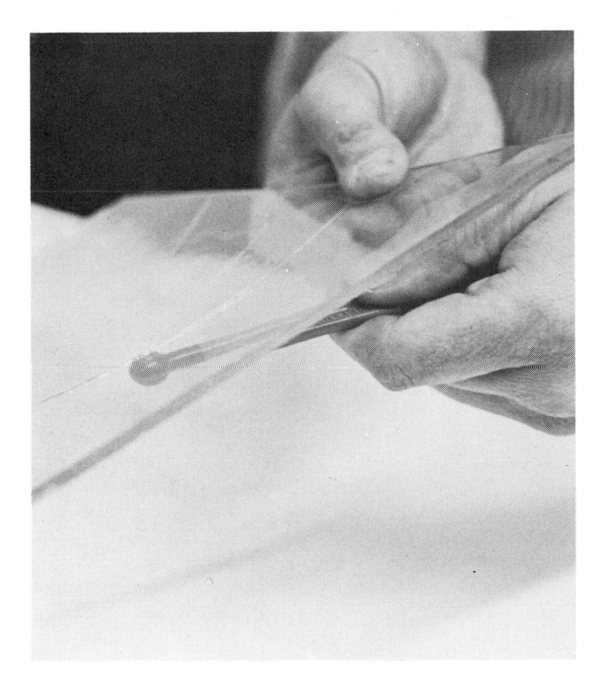

Tapping: the ball-end of the glass cutter is used to tap directly under and along the score line

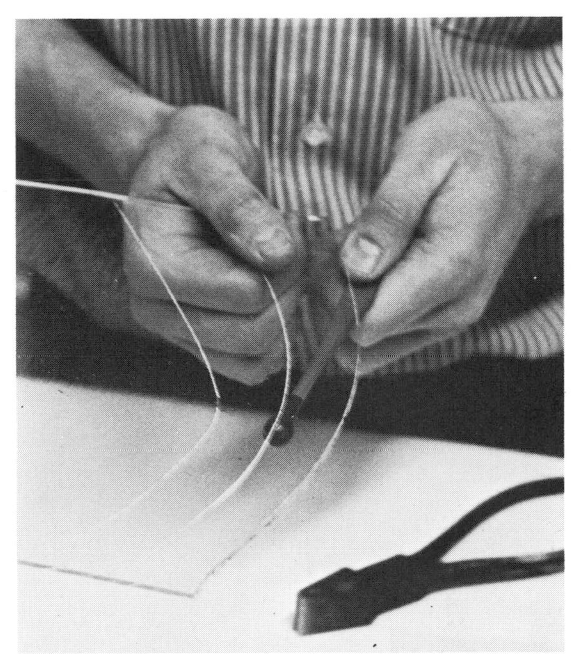

Both sides of the score-line are held firmly during the tapping operation

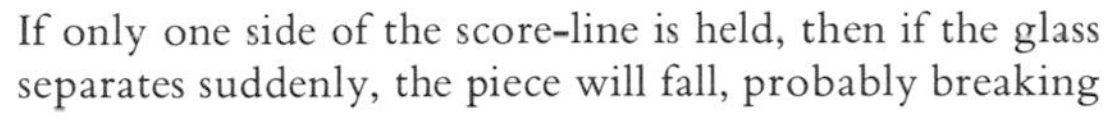

If only one side of the score-line is held, then if the glass separates suddenly, the piece will fall, probably breaking

Removing the scored and tapped piece of glass with glass pliers

One method of breaking a piece of scored glass is to use the edge of the work surface as a fulcrum. This can only really be used for larger pieces that entail straight scores to be broken, and is used primarily to break workable, handy sizes of glass off larger sheets. The sheet of glass is first scored and then positioned on the edge of the work bench or table so that the score is parallel to, and directly over, the edge. Downward pressure is then applied to the jutting out section with one hand (the other is holding the rest of the sheet firmly on the table top) and this movement should snap the glass directly along the score. It is a fast, nearly error-free method of breaking glass as maximum pressure is brought to bear along the score. Where, however, the piece of glass is too narrow to get a grip on both or either side of the score, then glass pliers should be used. Not only can one not exert enough leverage on a narrow piece of glass by hand, but whatever pressure that can be applied will be uneven, and the glass will snap off before running the entire length of the score. There is, in addition, the likelihood that such an event will cause the piece of glass, on suddenly snapping, to bite back into your hand. Alternative methods of breaking large pieces, and ones which glass dealers often use, both effectively and expeditiously, is to use cut-spreading pliers or to place the scored piece of glass on top of, say, a ruler placed on the work surface. The score is positioned along one edge of the ruler and a sharp, downwards pressure on the score should give a clean break along the score. If it does not, check whether you positioned the score directly over the ruler or whether a crack already existed in the glass which caused the incorrect break.

Which of the above methods will prove most suitable will depend on such factors as the size of the piece of glass to be cut, and whether the required score is straight or curved. The rule is that the straight-line scoring technique is used to divide a sheet of glass up into smaller, workable pieces, and that the final shaping of each piece – witha pattern piece – is done by means of tapping, glass pliers, and grozers. This is, though, a generalization as the method of cutting depends on the size of the glass you are using and on the shape of the pattern piece. You will, with practice, soon be able to determine which method of cutting is most suited to the particular job at hand.

Glass frequently leaves a jagged edge when it breaks, and to remove this, so that the lead came or copper foil can be placed flush against it, and also so that you avoid getting nicked by these razor-sharp edges, take the grozers and remove the splinters of glass by chewing them away. (If you don't have grozers then glass pliers are a good substitute.)

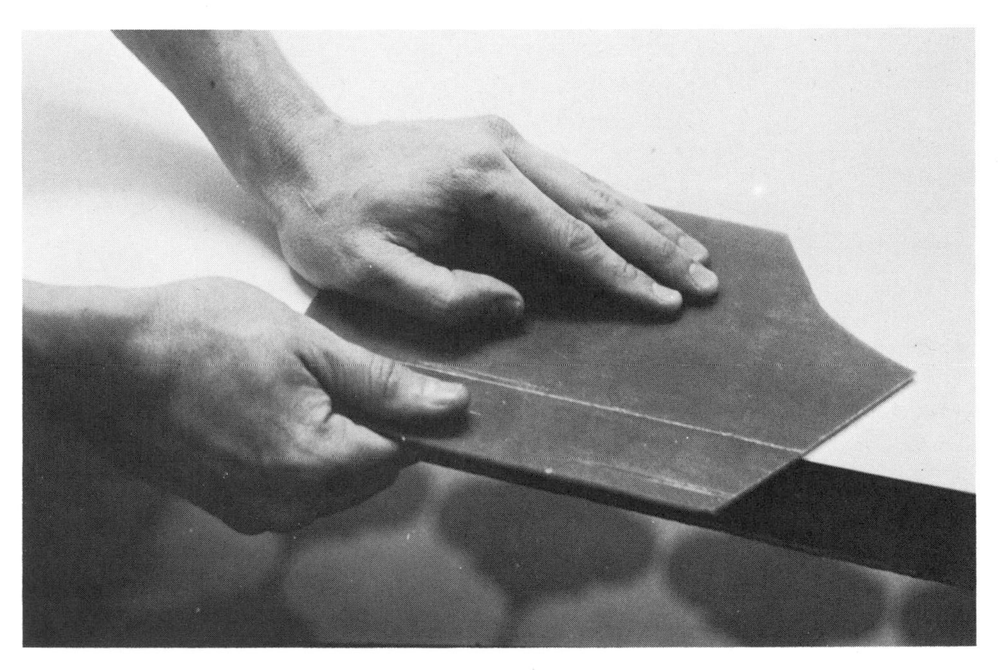

The straight score-line is positioned along the edge of the work bench

A sharp, downwards movement snaps off the extended piece of glass

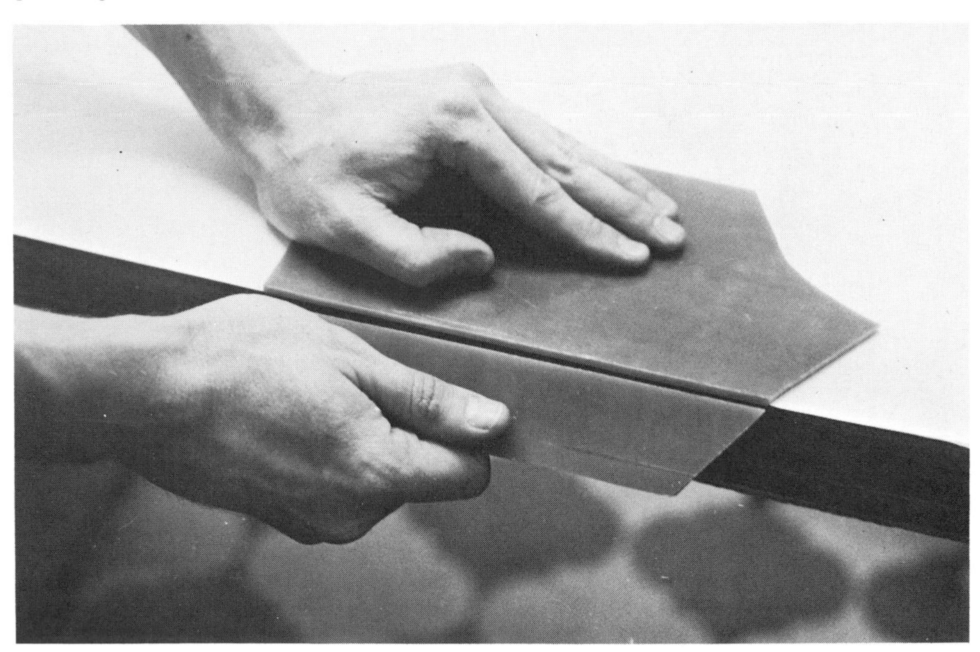

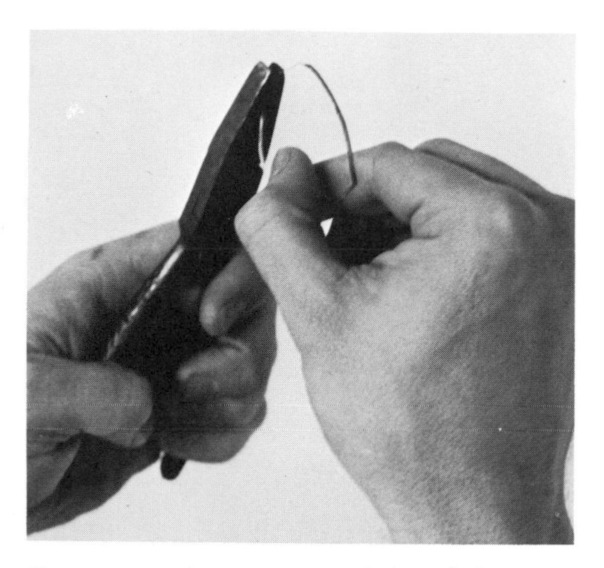

Use grozers or nippers to remove the jagged edges

The break in the glass has run off the score-line

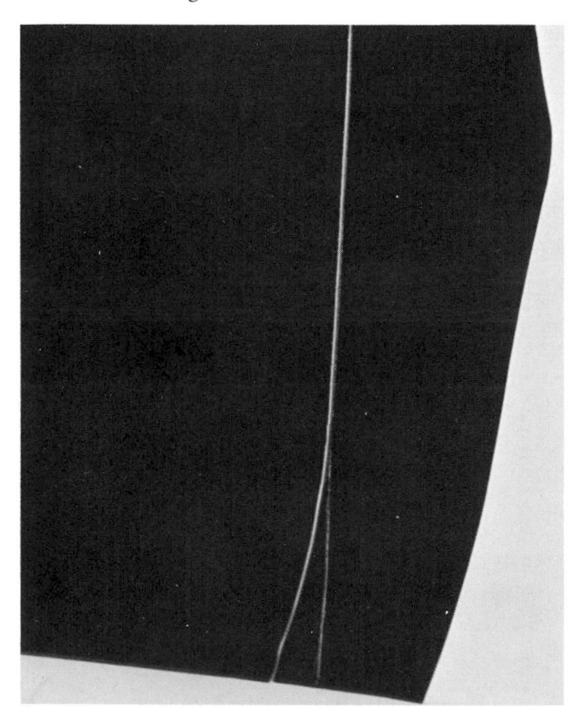

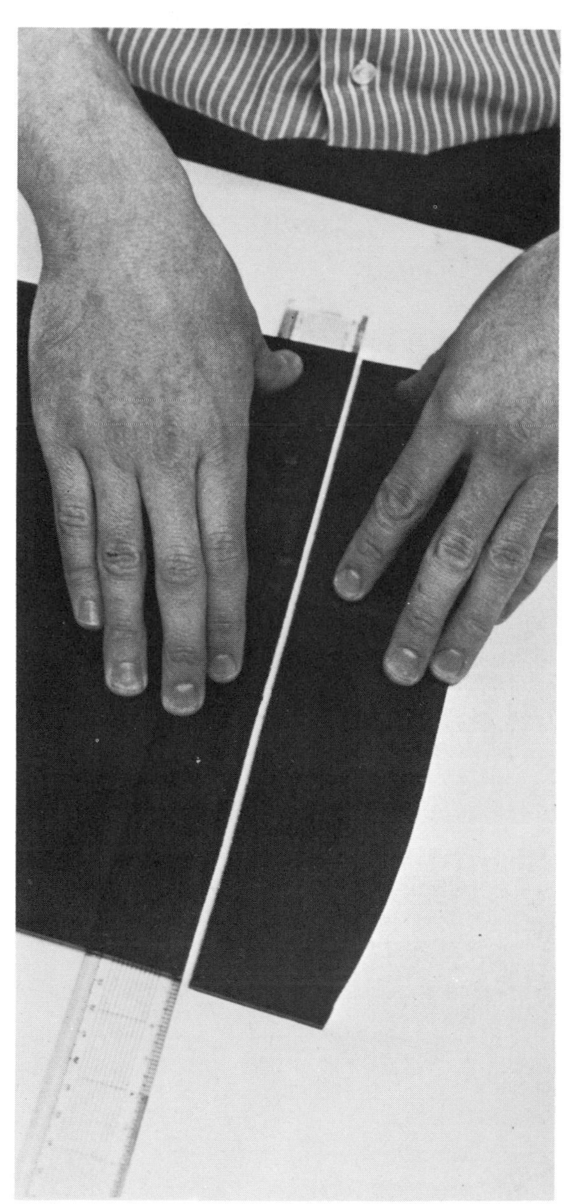

The glass breaks along the score-line placed directly over the edge of the ruler beneath it

Sharp convex and concave curves

Convex curves

If section ABC is to be cut out of the piece of glass in diagram (i), then first make the score ADE by cutting from A to D around that edge of the curve (you can save the large piece of glass on the left for later use). Then run the score off at a tangent to E. Next, tap the score in the same direction, from A to D, and, either by exerting gradual outward and downward pressure on both sides of the score with your hands or by using the glass pliers, separate the piece of glass on the left. Follow this score with another one from D to G, and likewise on to F, and then from G to C. Finally, round off any rough edges with the grozers, and this completes the convex curve.

When cutting a piece of glass out of a sheet try to minimize wastage. Economy in glass usage is something that you will soon develop: not only can glass be expensive, but when a particular sheet has been used it is invariably impossible to match it exactly for hue, striations, and density, with another sheet. So the more pieces that can be cut from a sheet the better. Select each piece from the sheet to get the best balance of glass economy and textural harmony.

Concave curves

These are the most difficult. Glass is most likely to break at its weakest point, and so it is often necessary to make several scores in order to reduce the pressure where such a break is likely to occur. For example, take the same shape as that used to produce the convex piece in figure (i), but this time it is the *other* piece, ACDEF, that is required. If you were to make a single score, ABC, and then try to break the glass along that line – either by tapping it with the ball-end of the glass-cutter or by using nippers – then in all probability the tap line would run off the score, and the sheet would break incorrectly. To lessen the pressure on the weakest part of the glass – in this instance, between B and G – the best method is to make a series of concentric scores parallel to ABC, decreasing in size. These scores do not, of course, have to be equi-distant. The major consideration is to remove section ABC gradually and in small pieces so that minimum pressure is brought to bear on the area between B and G. The more scores you make, therefore, the more likelihood of success. (There is, incidentally, no need to use concentric scores; an alternative method is to use a criss-cross/grill pattern of scores.)

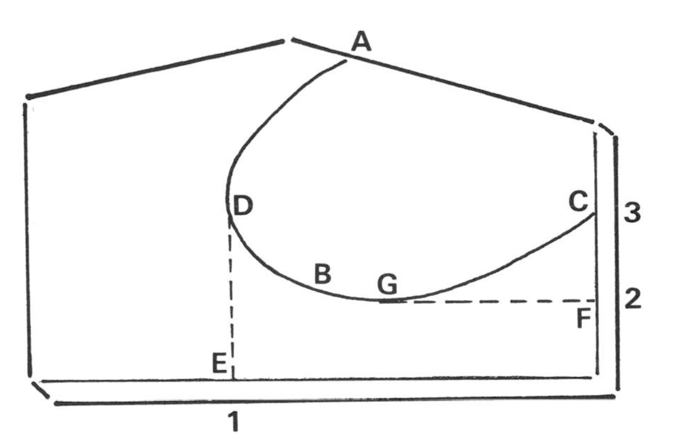

(i) Procedure for cutting out a convex piece of glass

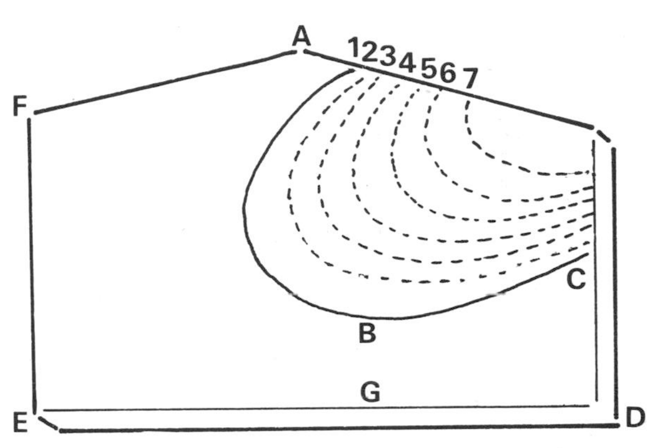

(ii) Procedure for removing a concave section from a piece of glass

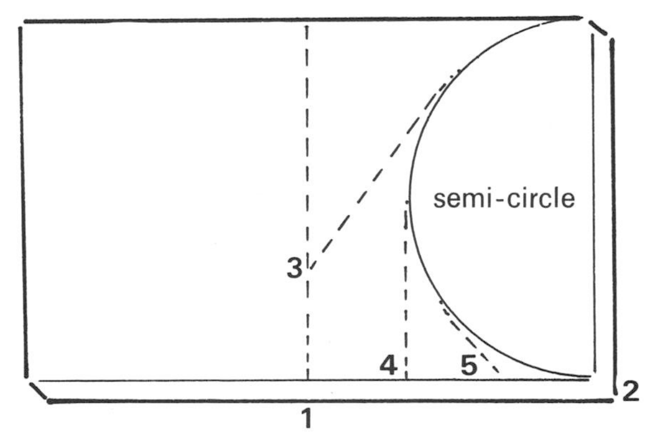

(iii) Procedure for cutting out a semi-circle. The numbers show the sequence of cuts

Proceed with more and more care the deeper into the curve you get, ie, the closer to the area around BC. When all of the inside of ABC has been removed, or even before if you feel that any more pressure from tapping or from the nipper will precipitate a break, then groze the inside edge of ABC to reach the score-line and to smooth any small jagged bits of glass that may remain.

With practice it will be discovered what can, and what cannot, be achieved with acutely angled concave or convex curves. The more proficiency acquired in this the more sophisticated the designs can be. Certain angles, however, cannot be achieved in glass, and in such instances an additional lead(s) must be added if you want to retain that shape within the design. Whether such additional leading does, or does not, detract from a design is a moot point and depends, of course, on the specific circumstances. In general, the less a piece of glass of the same colour has to be cut (and therefore the less leading needed), the more fluidity achieved in the composition. This is something that is very important to keep in mind when designing. Try to avoid angles which will necessitate leading *within* a *single* colour as this breaks up the rhythm. By all means use more leading *between* colours, ie, do not be dissuaded from attempting intricate designs such as an arabesque effect with leaves in a floral design so as to enhance the overall aestheticism, but unnecessary leading gives a cluttered and disharmonious effect. It breaks up a monochrome and makes it look as if you were unable to cut it all from one piece. This is probably the very reason why you have had to do it, and it will be apparent. The problem of attempting to eliminate such defects in the design cannot, however, usually be undertaken at this stage. It is too late. It is during the initial drafting of the design that such aspects of the actual materials with which you are working must be considered. You must constantly remind yourself of the limits of the intended medium. Ask yourself, 'Can this be done in glass?'

The above is a good guideline, and does hold true in most cases. There are, though, occasions when leading can increase, rather than diminish, the desired effect, helping you to obtain shapes that you could not otherwise cut out of one piece of glass, and enhancing the overall effect (see page 95).

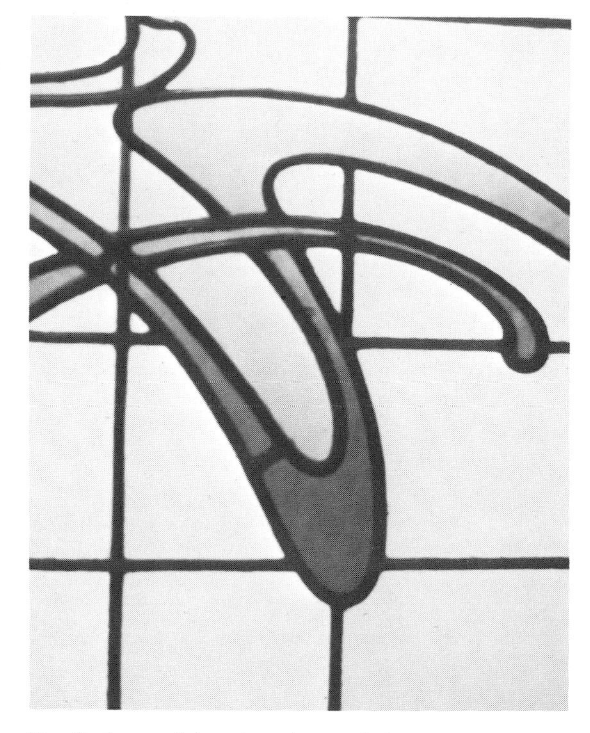

Detail of a panel showing advanced glass-cutting ability. One piece in particular, not for beginners

Cartoon sketch for a restaurant window showing how the design is broken down so as to avoid any impossible glass cuts. Notice, in particular, the neat, rhythmic manner in which the letters are subdivided
June Armstrong, London

Detail from a design for a leaded glass window. Not only does the introduction of leading (dotted lines) give a more natural look to each leaf, but it would have been impossible to cut five-lobed leaves such as these (maples) from a single piece of glass. Leading does, therefore, in this instance both increase the aestheticism and allow what would otherwise be an impossible shape to be incorporated in the design

The circle cut

Manufacturers of circle cutters claim two functions for their product: that it can cut both discs and holes out of glass. There is no question of their being able to cut discs, but do not be too optimistic about their being able to cut holes in glass. It is easier said than done.

(a) *Cutting discs*

Set the arm on the cutter to the required radius of the circle that is to be cut, and then, pressing down firmly on the central base of the cutter to prevent it slipping (it has a suction pad on the bottom for this purpose), inscribe the circle on the glass. Make the score line with a single cut if possible, because if it is scored in sections then the score is not likely to be continuous, with the end of one arc not running into the beginning of the next. This, in turn, can lead to an incorrect fracture when the glass is subsequently broken. Next, run a series of tangential cuts off the score line and tap these in sequence. Make as many of these tangential cuts as you like, as the more there are, the less acute the angle on the outside of the circle, and the more likely, therefore, that the glass will break along the edge of the circle, and not through it, like a chord.

When satisfied that the tapping has weakened the scores sufficiently, then remove each one with the glass pliers. Finally, groze away any corners on the circle.

(b) *Cutting holes*

The procedure is to first make the circumferential cut as before, and then to make a series of cuts inside the circle so that there is a grid of score lines. Do not, and this is vital, let the score lines run beyond the outside score line of the circle, as this will cause a certain break in the glass outside the hole.

Then tap the back of the glass along the outside line till it has fractured right round the circle. Now put the piece of glass down on the work bench so that the scored side of the glass is underneath, and then strike the centre of the circle lightly until it breaks.

Finally, use the grozing teeth on a glass cutter or offset pliers (see page 41) to carefully break out a channel in the glass before removing the rest of the inside of the circle.

Cutting holes is a hazardous undertaking. One needs both patience and luck to prevent the break in the glass from running off the circumference of the circle. If you try to do it, avoid very thin glass and allow at least 3 in. of glass all round the circle to give structural support to the tapping and breaking operations.

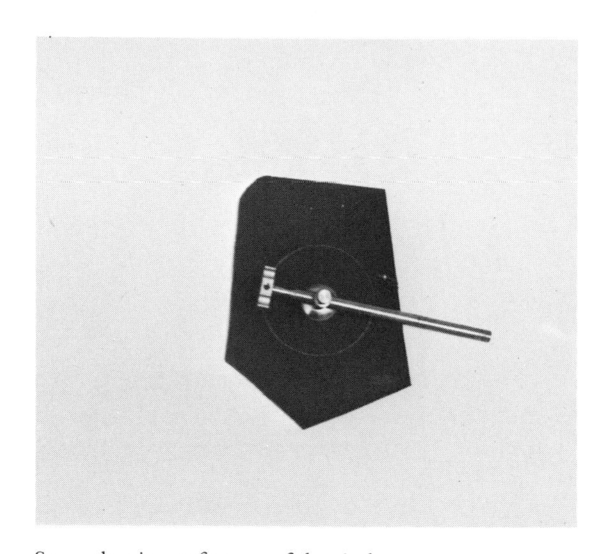

Score the circumference of the circle

Run the tangential cuts off the circle

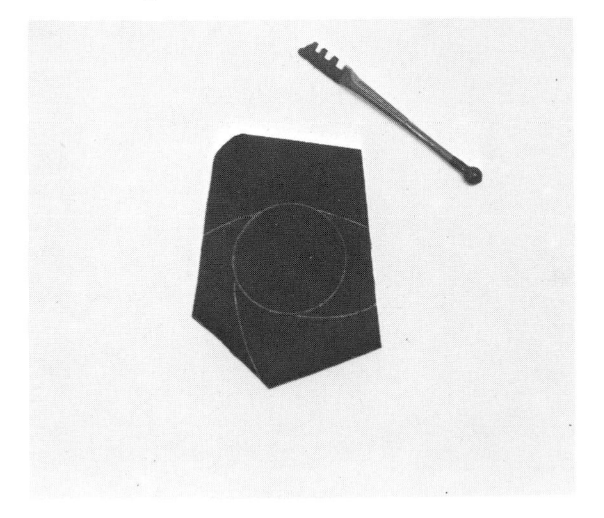

The circle after tapping

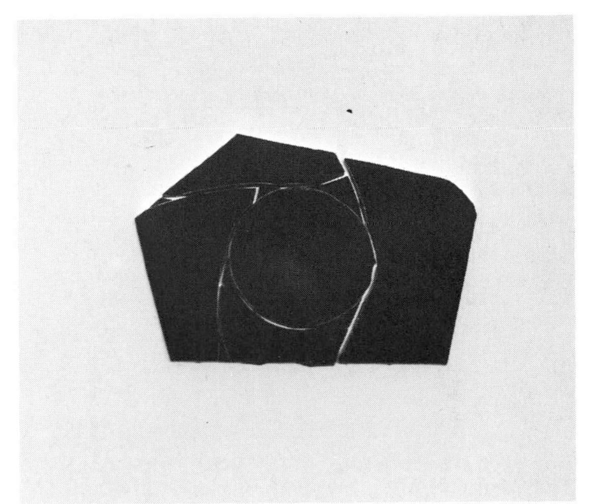

After tapping, remove the outside segments with glass pliers, and then groze away the remaining edges

The completed circle

Which side of the glass to cut?

There are no hard and fast rules on which side of a sheet of glass to cut – one is guided solely by pragmatism: by which side produces the cleaner, more predictable, separation of the glass.

Certain proven guidelines do, though, exist. First, on most varieties of glass, such as American machine-rolled ones, one side is smoother and more shiny than the other, duller, side. Which is which can easily be determined by holding the glass at an angle so that the light reflects off the glass' surface. The smoother side is often, especially for beginners, considerably more difficult to score as the cutter tends to skid off it or not to be able to keep exactly to its intended course. The surface can, simply, be too slippery, leading to a broken score line and a subsequently bad break. If you find that this is the case then cut on the duller side where the cutter will get a more substantial grip on the glass. (On glass which has a very pronounced surface texture, such as a nodular pattern, you can, of course, only cut on the smooth, ie, non-textured, side.)

Secondly, when using flashed glass (glass which has a thin layer of coloured glass fused on to the thicker base colour, which is usually white or clear, for example, flashed ruby on white), then always cut on the *non*-flashed, base-colour, side. This is because one cannot always get a clean cut on the flashed side, which can, in turn, lead to the glass splintering or shattering when it is separated.

Thirdly, opalescent glass can be considerably more difficult to cut, whichever side you try, than transparent glass. This is because the internal pressures generated during manufacture by the different coefficients of expansion in each oxide lead to unpredictable breaks.

The above guidelines can, however, only help to facilitate the cutting process if the wheel of your cutter is sharp and well-maintained: a blunt wheel makes it equally difficult to cut a sheet of glass on whichever side you try.

Chapter 6

Project in glass

When the methods of cutting glass and working with lead came have been mastered, these can be incorporated into an actual glass project.

A good method by which to illustrate the entire process – one which is widely used as part of an introductory course in leaded glass work by instructors in America – is to produce a leaded glass panel. First a design which is suitable must be selected. This will be transferred on to a cartoon and from there, finally, translated into glass.

Circular panels are more difficult to assemble than rectangular ones, and should not, for this reason, be attempted by beginners. In this panel Buddhism comes to glass in the form of a Tantric Mandala
Kathy Smith, Berkeley, California

The design

This is not yet the time to give free rein to your artistic ability. No, in this introductory stage you will have to bridle your creative talents and concentrate on producing a relatively simple composition. This is, however, in no way intended to take the fun out of it. Any number of simple and eye-catching designs can be created by using basic geometric shapes. A combination of such when combined with a well-balanced choice of different coloured and textured glass, can result in a panel that is both symmetrically and aesthetically pleasing.

The object at this stage is to become familiar with the various materials and instruments of the craft, and in so doing to realize their decorative and constructive potential – and limits. Your *tours de force* will come later, but not until you have perfected this initial stage. So bear with it.

Take a sketch pad and work out a design. Concentrate on producing an end-product that can be cut out of glass; certain shapes that are easily drawn on paper cannot be achieved in glass. It is easy enough to cut a pattern out of paper, but it may not be possible to do the same in glass, or, if not impossible, then may produce difficulties in the leading stage that will detract from the intended effect.

Here are some examples of the kinds of problem that can arise, but there are others that will be recognizable at the design stage, and so possible to prevent. For example, if the design consists of several triangular shapes converging into a single point, then the conciseness of the effect on paper

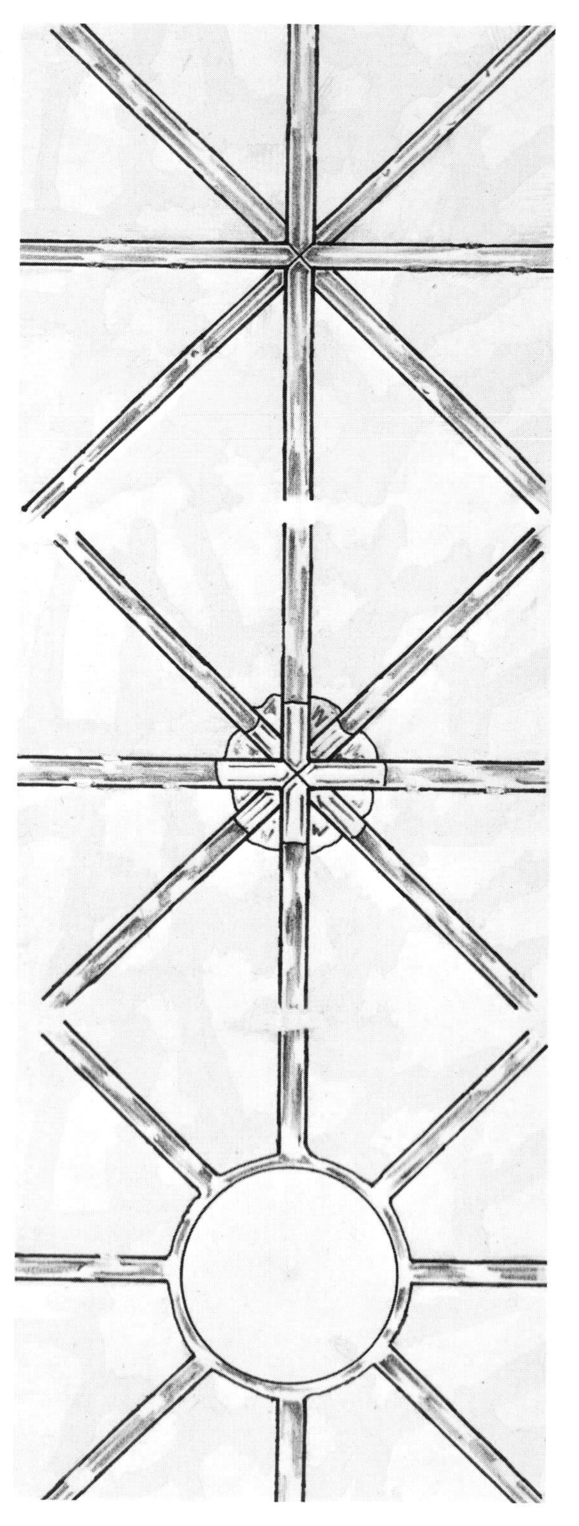

The intended effect of a design that consists of several leads meeting in a single point.

After soldering, however, the sharpness of the design is lost as the solder overlaps the glass

A suggested way of overcoming this problem: the centre can consist of either a circular piece of glass or of a glass jewel

will be lost in leading the glass as all the leads meeting at the same point must be soldered, and the solder will probably overlap the glass no matter how carefully you have mitred the lead endings. The result will be a messy blotch of solder over the central joint rather than the neat welding together of the leads that had been intended. It will look slapdash. Such a design will, in addition, be structurally weak as the concentration of leads in one place will put too much stress here – pressure at this point would be likely to force the whole panel to break outwards.

Be careful, also, in using a long, narrow piece of glass as the leaves of the lead cames on both sides of the piece will overlap the glass surface to the extent that the glass may be 'filled in' so that none of it remains visible. The result will be entirely different from what had been intended; instead of a clearly defined area of glass bordered by two strips of lead, one is left with a 'double-leaded' effect that visually swallows up the glass and which may consequently be far too heavy for the composition thereby disrupting the overall symmetrical balance. So try to anticipate where this might arise in the design. There are ways of circumventing this problem, such as incorporating thinner leads so that more of the glass is visible, thereby making the presence of the lead less pronounced (the technique of varying lead widths for artistic effect is discussed in the next chapter), but at this stage we will use only one width of lead and so must concern ourselves more with the prevention of such a situation, rather than with its cure.

Remember, also, that until you are fully aware of what you can, and cannot, achieve in glass, very sharp concave angles should be avoided.

It is sometimes a good idea, especially in this initial project, to paint the sketch so as to achieve the optimum colour balance. Any colouring device can be used: coloured pencils, felt-tipped pens, water paints, or oils. Pastel water colours are probably the best as they usually give the closest approximation of a stained glass effect. This method of obtaining a colour balance, or of minimizing the possibility of a glaring imbalance, is recommended only if you do not have access to a light table. With the latter, shuffle differently coloured pieces of glass around on the top of the table till the optimum effect is achieved. This will enable you to see what the end result will be like, because the panel will, finally, be placed against a light source, whether natural or artificial. What invariably happens is that the colour scheme first decided on is not the one that you eventually choose – while interchanging the pieces of glass you often find a better one.

The size and shape of the panel itself can, with one constraint, be left to you. This is that it must be right-angled, ie, square or rectangular. (More ambitious shapes such as circles, triangles, rhomboids, and polygons, can come at a later stage.)

Introductory classes in the technique of leaded glass often recommend a specific size, say, a 15 in. square panel. This does in itself serve very adequately from a technical standpoint, but the end result is that one has a panel which one cannot put to any particular use. It is not functional, unless, by sheer circumstance, one happens to have a 15 in. square window space into which it can be fitted. So, before you begin, see if there is a small existing window which you would like to replace with your own one, and then measure it up and use these dimensions for your panel. If, however, there isn't one, there are other ways in which the finished product can be displayed. Suggestions will be offered at the end of the chapter, but everything you make in stained glass – even this initial project – should be put to use. There is no point in utilizing time, money on materials, and your artistic talent, to no avail.

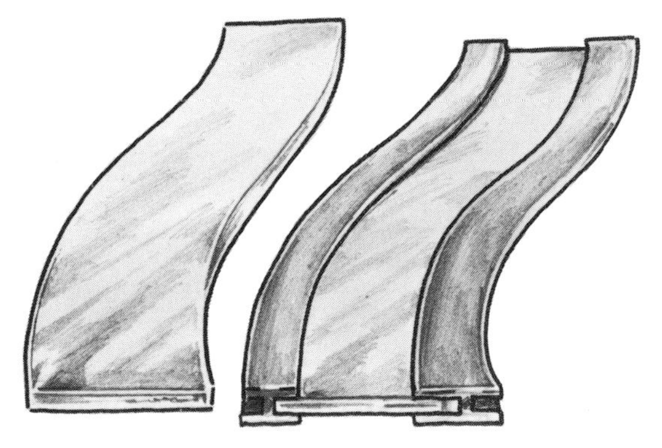

The leaves of the lead came overlap the surface of the glass, thereby reducing the amount of glass that is visible

The dimensions of the perimeter of the panel depend on the channel measurements of the border lead, in this case ½ in. H-frame lead

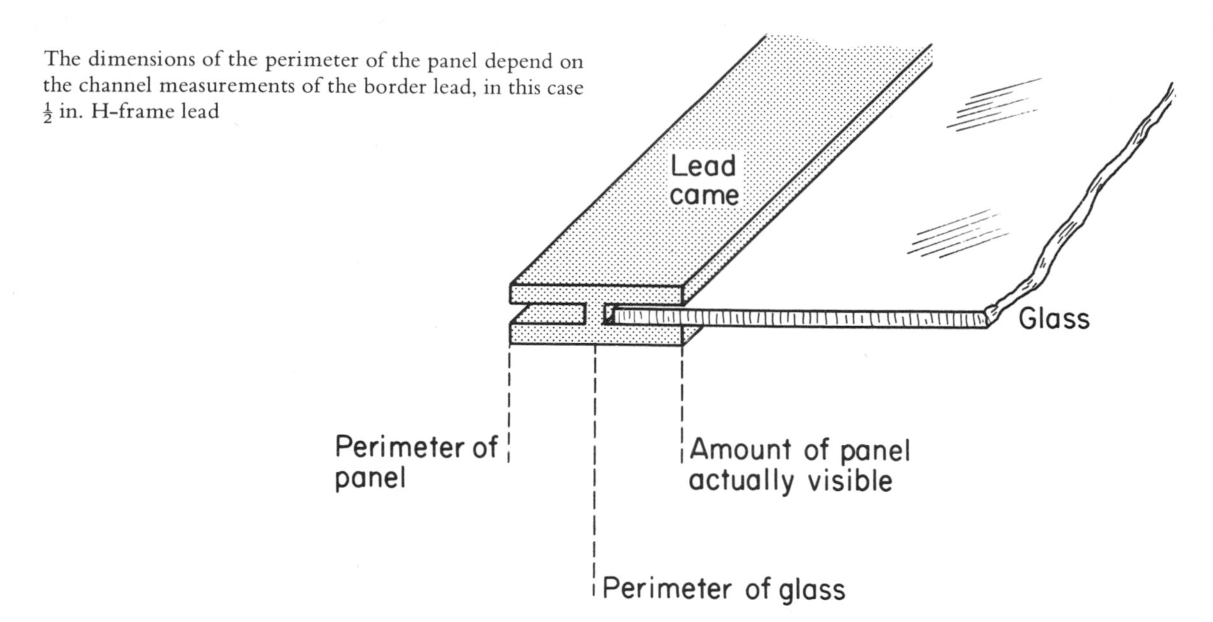

The following is a design that I have chosen, and the one I will use to show the various, sequential steps from sketch to finished panel. It is by no means an artistic extravaganza, but will serve to show the rudiments of working with leaded glass. You will probably prefer your own design, and if this is the case, then use it. But if you are stuck for one of your own – and it is amazing how often one's mind 'freezes' when one first tries to be creative; what one pictures in one's mind just doesn't 'happen' when transferred on to paper – then follow the design on page 103.

Do not, incidentally, if you have your own one, be dissuaded from using it because of any misgivings that you would not be able to transfer the instructions for the design in this chapter into your own panel. You would. The directions given here are procedural in a general sense and apply to any panel or window. The only things that change are the shape of the pieces of glass and the length of the leads; the technical aspects of using a cartoon, lathekin, nails, etc, remain exactly the same whatever the design.

When the design has been settled on, then the next step is to make 3 copies of it, the function of each being as follows:

1 *The cartoon paper copy*
The design is drawn on the cartoon paper which is used, in its turn, to make the additional work drawing and pattern paper copies of the design by means of carbon paper.

The cartoon is used during the assembly of the glass panel as a working guide, and can be placed on the work surface alongside the panel or be pinned up in front of you for easy reference. (You can eliminate one of the three copies and use only the work drawing and pattern paper. In this case the cartoon becomes the work drawing, and the extra copy, used as a guide, is no longer used. I recommend, however, that you use all three copies: this acts as a safeguard in the event that the copy being used as the work drawing gets torn or otherwise damaged. You will, then, have another copy of the design to which to refer.)

2 *The pattern paper copy*
For this use a heavier texture than the cartoon paper. An office file folder is, for example, of an ideal thickness and strength. The reason that it must be firm is that it is used to form the individual pattern pieces, ie, templates, with which each section of glass in the panel is cut out. If it is too thin then it will not retain its shape when the glass cutter is run around it; the pressure from the cutter can cause it to buckle, and the resulting piece of glass will not have the correct dimensions, creating problems when you add the leading later. At the other extreme it must also not be too thick as it must be thin enough to fit between the cutting edge of the wheel and the axle on the body of the cutter itself.

The design for the panel to be assembled in this chapter

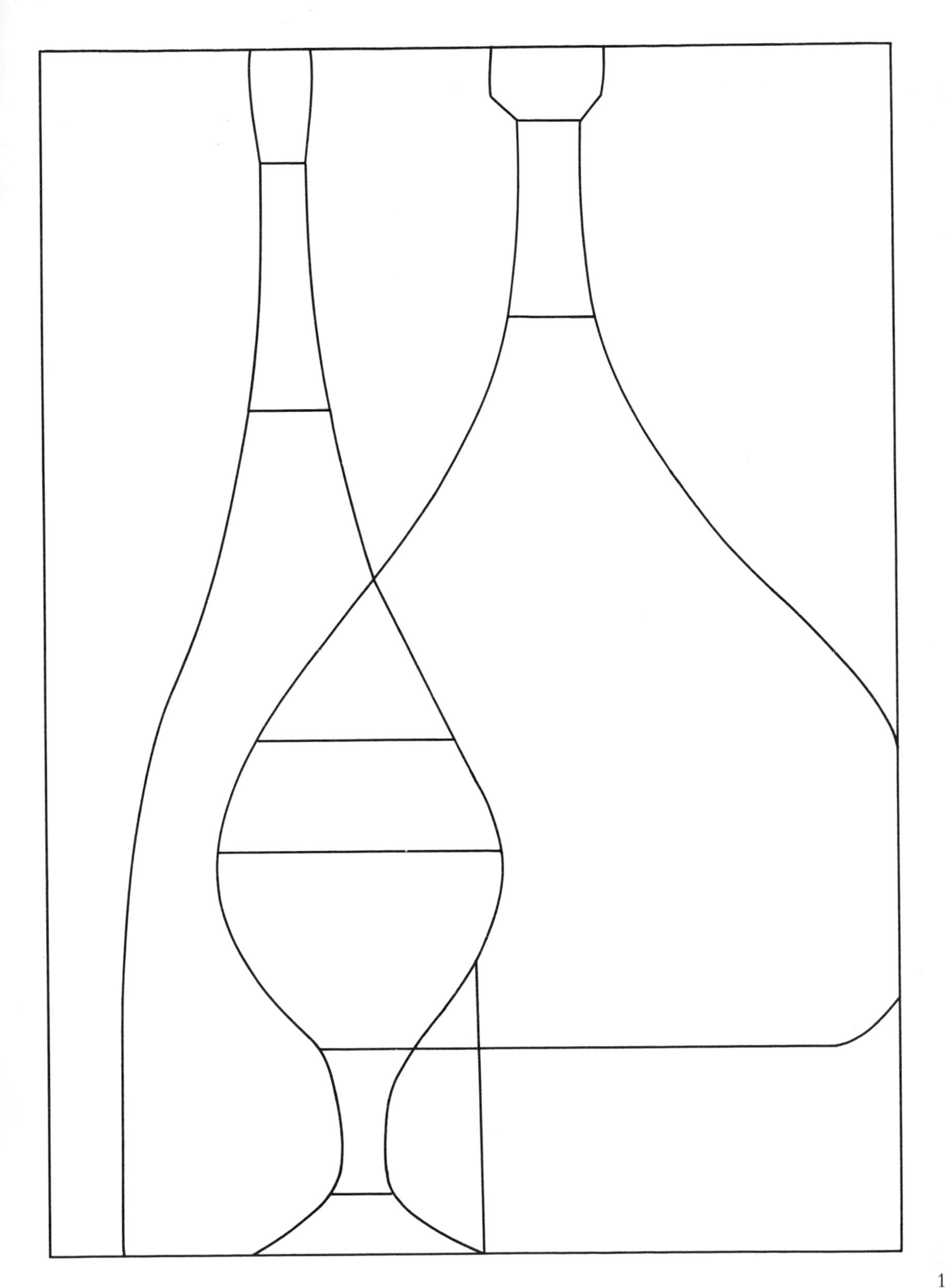

A pattern piece does not usually last very long: the abrasive effect of the cutter pressing down along its edge tends to fray or corrugate it, and if you are dipping the cutter frequently in a cleaning fluid (such as kerosene), this has a further corrosive effect on the paper's edge. If, therefore, you have to cut multiple pieces of glass of the same shape, then it is advisable to check the pattern piece regularly to ensure that it is retaining its intended dimensions. Cut a new pattern piece when it shows wear.

3 *The work drawing copy*

This should be a fairly sturdy *kraft* or cartridge paper which is placed on the work surface so that the glass panel can be assembled on top of it, piece by piece. If it is too thin then it is likely to tear or crinkle. Both *kraft* and cartridge are supplied in roll or sheet form by paper stockists. Buy it in sheets at this stage, but if you find that you are using large quantities then there comes a break-even point after which it becomes economical to buy it in rolls.

Now take the cartoon paper and draw a 15 in. × 10 in. rectangle on it. To make sure that it is right-angled I recommend the use of a set-square. When you have done this draw another rectangle $14\frac{1}{2}$ in. × $9\frac{1}{2}$ in. in the middle of the first one so that there is a $\frac{1}{4}$ in. margin all round between the two perimeters. Before you can understand the reason for these measurements it is necessary to know the sizes of the leads that will be used in the panel: two different sizes of lead will be used – a $\frac{1}{2}$ in. flat H-lead for the outside border, and a $\frac{1}{4}$ in. round H-lead for the inside leading. As lead is supplied in lengths of 5 ft, so two strips of the former and two of the latter should be sufficient for the project.

The perimeter of the panel is made up of a $\frac{1}{2}$ in. H-lead, and this is placed so that its outer edge lies along the 15 in. × 10 in. line. This means that one half of the H-lead, ie, the $\frac{1}{4}$ in. from the outer edge of the leaf to the heart, is not actually going to be used. It remains empty, in fact. So, because of this $\frac{1}{4}$ in. border right round the panel that is not being utilized, so a total of $\frac{1}{2}$ in. – $\frac{1}{4}$ in. on each side – must be subtracted from the length and breadth of the panel to mark the actual area that the glass will take

Transfer the original sketch on to the cartoon paper, making certain that it has adhered to the required proportions

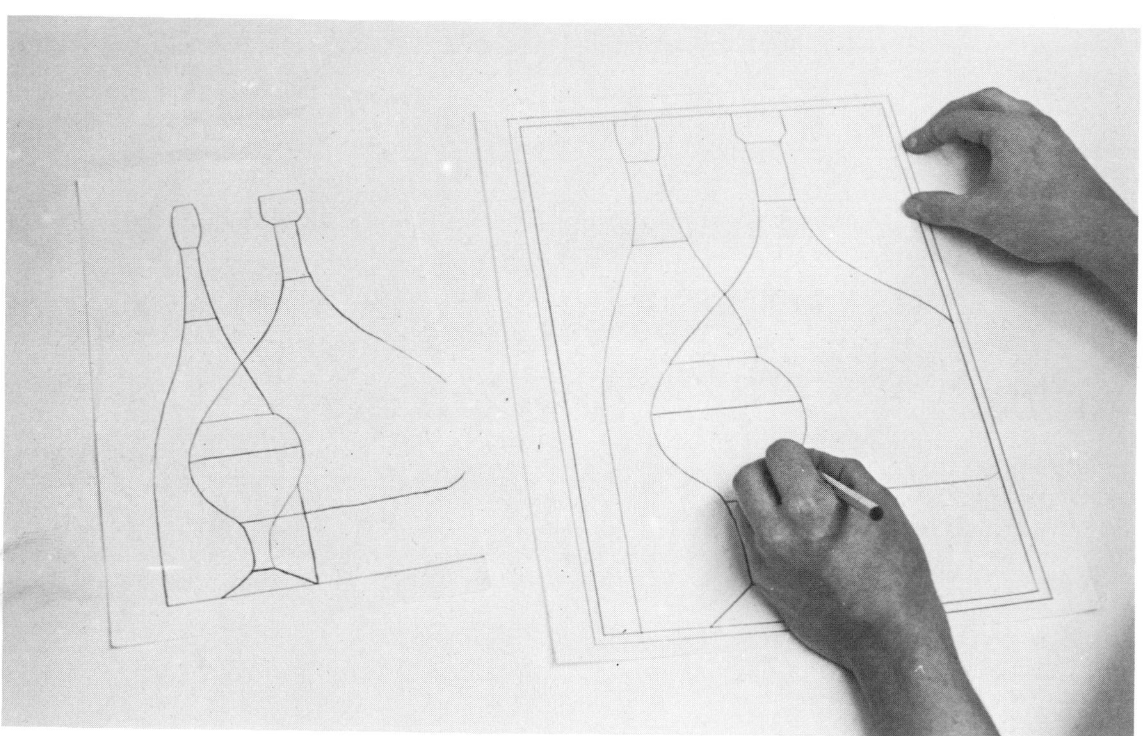

up. This, then, is the inner 14½ in. × 9½ in. rectangle. In addition, a further ¼ in. of the design will not be visible as it is hidden by the inside ¼ in. leaf of the H-lead, so the actual visible size of the panel is 14 in. × 9 in.

Perimeter leads: H or U frame?

Professional stained glass craftsmen in England always use an H-frame for the outside of a panel or window. This is because the outer leaf will not show under the moulding or flange of the window frame, and so this empty channel allows a certain margin for error should the window be too large. In such circumstances, the outer channel is either mitred down or bent inwards to reduce the overall dimensions of the window, thereby enabling it to fit.

It is probably because of this, ie, that stained glass work has traditionally been incorporated into wooden and steel frames (thereby ensuring that the outer channel of the lead is never visible), that the need for a U-frame lead has never materialized in England.

It is for both of these reasons that I have used an H-frame as the perimeter lead in the panel in this chapter: firstly, in the event that the panel may be used as a window, and, secondly, that it may not be possible for you to obtain strips of U-frame. Remember, though, that if you intend to display the panel so that the outside border is visible, then a U-frame should be used or else the outside channel of the H-frame – whether left empty or even mitred down – gives an unfinished and amateurish appearance.

The next step is to transfer your original sketch (design) on to the cartoon paper so that it fits into the 14½ in. × 9½ in. rectangle. Make certain that the design is in the same proportions as the original sketch. If you don't feel confident to do this free-hand without distorting the design, then divide the sketch up into equal sections by means of a grid pattern, and then, using the same grid (adjust the size proportionately if need be, on the cartoon paper), re-draw the lines in each section of the grid on the sketch into the corresponding section of the grid on the cartoon. This will minimize errors.

The next stage is to make the two additional copies of the design on the cartoon paper. Place the pattern paper down first as it is the thickest, then the work drawing paper above it, and then finally the cartoon on the top. Place a piece of carbon paper (face down) between each of the three sheets and then square them up. When you have done this then secure it all with drawing pins (thumb tacks) at each corner.

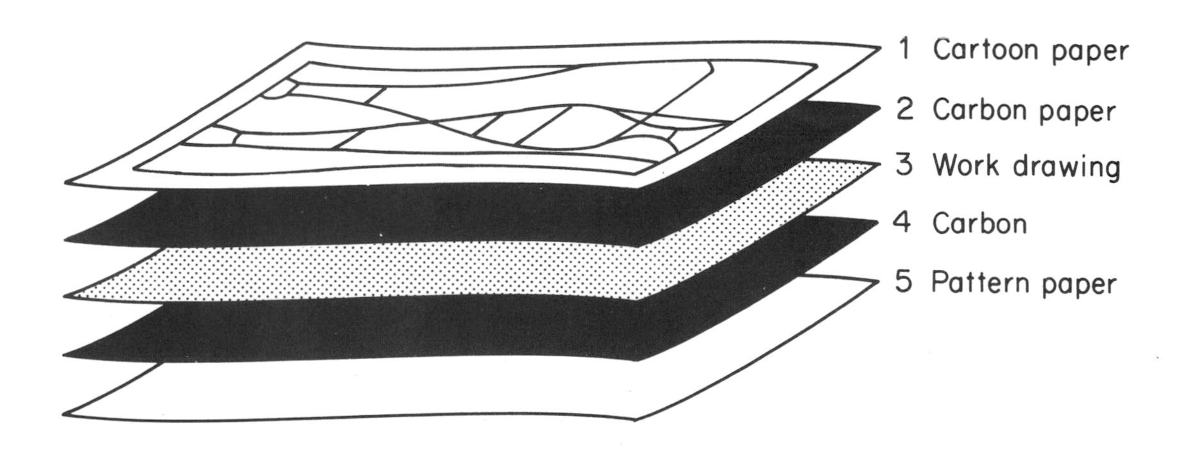

Trace over all the lines of the design on the cartoon paper, pressing down firmly to make certain that the design is transferred on to both of the lower sheets. Now number each piece on the design. It is advisable, in addition, in more intricate designs, to colour code each piece; for example 11B could indicate that the eleventh pattern piece is of brown glass, or that it is a background colour. Develop your own colour coding system. When completed, remove the two bottom drawing pins and check whether the design has, in fact, carried through on to the lower sheets. If so, then remove the remaining two drawing pins and put away the pieces of carbon paper. You are left with three copies of the layout for the panel. Next, take the pattern paper. Cut along all four edges of the inside perimeter line of the pattern (size $14\frac{1}{2}$ in. $\times$ $9\frac{1}{2}$ in.) with a standard household pair of scissors. The next step is to cut out each individual piece in the design. This, however, should not be done with standard scissors as allowance has to be made for the lead came that will lie between each piece of glass. The heart of the lead is $\frac{1}{16}$ in. wide, and so this must be allowed for when cutting out each pattern piece or else the total panel will be correspondingly larger than planned.

Pressing hard, trace over the design and then number the pieces in the order in which they will be assembled in the panel

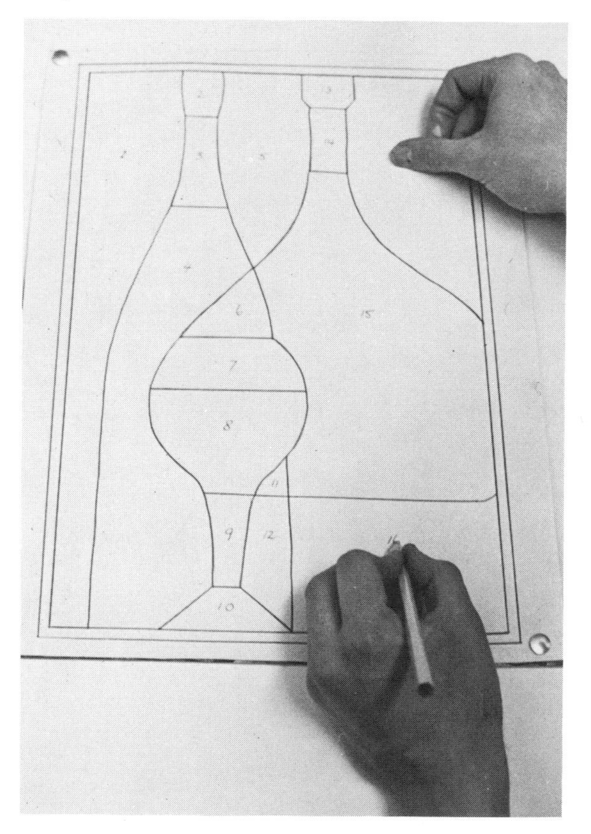

Remove the bottom two drawing pins (thumb tacks) to check whether the design on the cartoon paper has carried through on to the work drawing and pattern paper underneath

Pattern piece cutting

There are three methods of cutting out the inside lines of the pattern paper to ensure a $\frac{1}{16}$ in. gap for the heart of the lead.

1 *Three-bladed pattern shears* These are made for the job, ie, to cut out a $\frac{1}{16}$ in. wide strip of paper, but are not readily available in England. They are, in addition, expensive, but fortunately not essential. If, however, you have a pair of pattern shears, then place them at the beginning of a line so that the line falls between the two outside blades. Cut around each pattern piece. It is the inside blade, sandwiched between the two outer ones, that is removing the $\frac{1}{16}$ in. line. Use short strokes as the strip that the shears are cutting out tends to get snarled up and should be broken off every 1 in. or so. Hold the pattern paper in one hand close to where you are cutting and work the blades of the shears along the line with the other. Use even shorter strokes when cutting around corners as the shears are cumbersome and tend to jam up when negotiating angles.

2 *Razor blades* Take two single-edged razor blades and place a piece of $\frac{1}{16}$ in. cardboard or wood between them and then tape them together. This does the job very adequately and manages to cut round the corner on the pattern paper with far more manoeuvrability than the pattern shears. I recommend this method: cheap, fast, and functional.

3 *Pencil lines* Two alternative methods, but not recommended, as the width of the cuts will probably vary, thereby causing inaccuracy. One: draw a thick $\frac{1}{16}$ in. line around each piece on the pattern paper and then cut each pattern piece out with standard scissors along the outside of the line. Two: draw a double line $\frac{1}{16}$ in. apart, and likewise cut along both lines.

When all the pattern pieces have been cut, lay them on the cartoon paper in their correct positions. They should fit so that there is only the $\frac{1}{16}$ in. gap for the lead running as a border to each pattern piece. Check that this is, in fact, the case. It is the last chance to discover if there are any errors before you cut the glass.

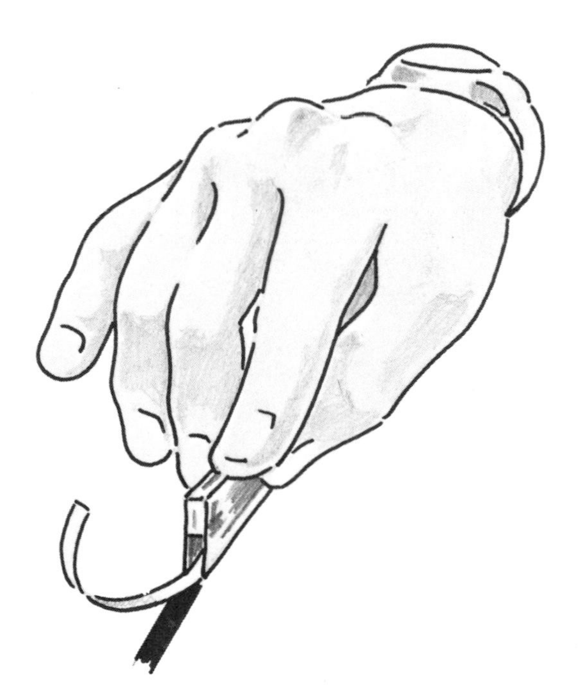

Using the razor blades to remove the $\frac{1}{16}$ in. required for the heart of the lead

Now you are ready to cut out each piece of glass. Take each pattern piece – the order does not matter – and, making certain that you are using the predetermined colour of glass for each section of the panel, then proceed to cut all the pieces. Use of a pattern piece involves exactly the same cutting techniques explained in chapter 5 with the advantage that the edge of the pattern piece is, literally, a guideline to ensure that the cut piece of glass exactly fits the dimensions required in the panel.

The following are useful hints to bear in mind when using a pattern piece:

1 Glass economy requires that as little glass as possible is wasted. Place the pattern piece near the edge and/or corner of the sheet and then cut away the remaining larger piece of the sheet to prevent it breaking while you are cutting the piece at hand. A small piece of glass is, in addition, much easier to handle.

Position the pattern piece near the corner of the sheet of glass and cut away the remaining, larger segment of glass before cutting out the required piece of glass by means of the pattern piece

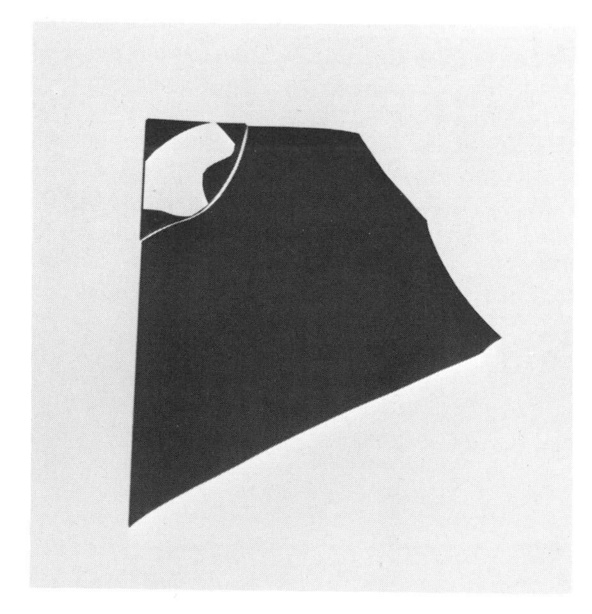

2 Hold the pattern piece down firmly with one hand while making the scores. It must not be allowed to slip even a fraction as this will result in an incorrectly cut piece of glass. This, in turn, means that the leading will likewise have to be adjusted, and the dimensions of the adjacent piece of glass may also need to be altered so as to allow for this initial error. Mistakes tend, in fact, to have a compounding effect: the first one requires an adjustment, which in turn requires another one, and so on. Sometimes a piece of glass that seems to have been correctly cut still does not quite match the dimensions of the pattern piece. The reason for this can be, if no slippage has occurred, that glass does not always break perpendicularly from the top surface to the bottom. It sometimes breaks at an angle, leaving the bottom edge slightly wider than the top. If this happens then turn the glass bottom-side up and gently groze away this overhang.

3 It is advisable to make the hardest cut first as once this has been achieved then there is less chance that the glass will break incorrectly while the remaining scores are being made. This is especially important where a required piece of glass includes a deep concave curve. It is most frustrating to complete, say, seven out of eight facets of a piece of glass, only to have the last one break incorrectly.

Using a pattern piece. The cutter follows the edge of the paper

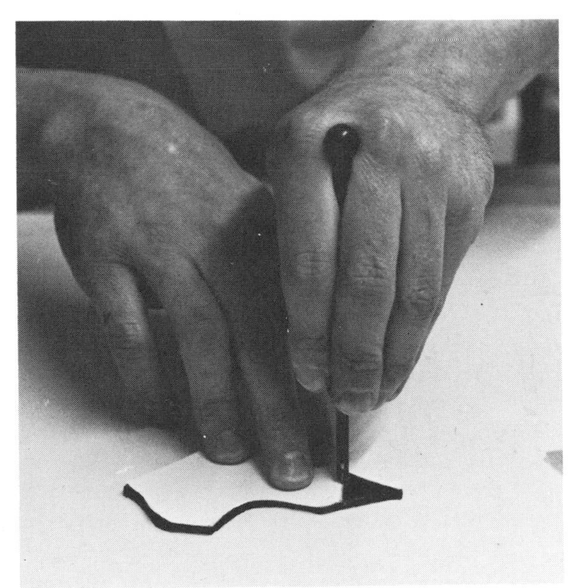

4 When the piece of glass has been cut, then place the pattern piece on top of it to ensure that the two match exactly; where the glass is slightly larger – usually on the corners – then groze away the excess glass till you are satisfied that they are identical. If, on the other hand, the pattern piece is larger than the glass, then it probably slipped while you were cutting, and the piece of glass will, therefore, have to be re-cut.

A word of caution about grozing. Its function is to supplement the scoring of the glass, not to replace it. It should, if one is cutting correctly, have to be used minimally to 'round off' any discrepancies between the size of the glass and the pattern piece. Use the technique sparingly; it should not be used excessively at the expense of one's cutting ability. It is, in addition, far more likely to lead to the glass breaking incorrectly than is a well-cut score.

(a) Glass sometimes breaks so that the bottom edge is wider than the top.
(b) Turn the glass over and gently groze away the overhang

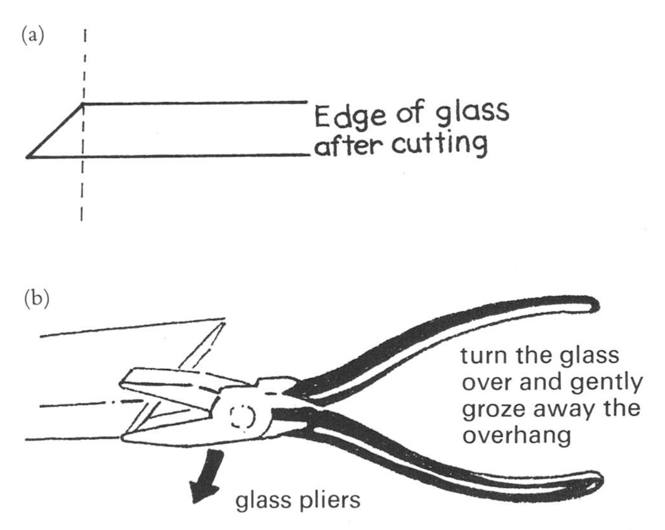

The traditional method of cutting out a piece of glass. The glass is placed on the cutline which is itself on top of a light table. The glass is scored freehand without the aid of a pattern piece
Reproduced by courtesy of Goddard and Gibbs, London

Pattern pieces are not used within the stained glass trade in England. The time-honoured method is to place a piece of glass on top of a cut-line which is itself on top of a light table. The cut-line is a thin tracing paper and the lines for the lead came in the design can be clearly seen on it through the piece of glass by means of the transmitted light from the light table. The cuts are then made free-hand on the glass, ie, without the use of a pattern piece. The absence of the pattern piece stage reduces the overall cutting time considerably, and most stained glass craftsmen would consider its use to be time-wasting and superfluous. There are two reasons, however, why I feel it is necessary for our purposes:

1 People cutting stained glass professionally are master craftsmen. Many of them have had a number of years of experience in the technique of scoring glass, and speed and dexterity in cutting are to them second nature, so they do not therefore need the guideline edge of a pattern piece. They also have all the ancillary equipment of their craft – in this instance I am referring specifically to light tables – at hand in the studio. You, however, may lack both of the above, neither having undergone a lengthy apprenticeship in glass cutting, nor perhaps having access to a light table. The use of pattern pieces overcomes both of these potentially limiting factors.

2 The second reason is that the glass used in England in professional stained glass work (mainly conventional church windows) is transparent. Even the darker antique sheets tend to transmit sufficient light for the lines on the cut-line beneath the sheet to be seen by means of the light table, and it is only in very rare instances that glass is used which is too dark to see through. When this does occur, one of the methods used to overcome this problem is to place the cut-line on top of the piece of glass to be cut (the reverse of their normal positions), and to place a piece of carbon paper face downwards between them. The outline of the glass is then traced on the cut-line, and this transfers a greasy carbon image of the cut-line on to the glass. This, in turn, is dusted so that the perimeter of the piece of glass will show, and then, finally, the glass is cut out along the perimeter as marked.

This is, though, obviously an expediency measure developed for infrequent use of very dark glass.

The craft of leaded glass, however, utilizes opalescent glass, and as this is, by definition, opaque, so it is impossible to determine where the lines are on a cut-line placed beneath it. So pattern pieces are essential.

Only one of the seventeen pieces of glass may, I think, cause problems in cutting, and so, in the event of difficulties, the procedure of cutting this piece is described.

This is the one that corresponds to pattern piece 4. It contains a concave edge on one side which should be cut in the same manner as that described for concave curves in chapter 4. The diagram shows the scores, in sequence, that were made to cut out piece 4 (the sheet from which your 4 is cut will, no doubt, be of a different shape; this does not matter: adjust the cuts for scores 1, 7 and 8 accordingly, adding, if necessary, cuts 10 and 11).

Proceed as follows:

1 The first score, 1, is to remove any excess glass to the right of the side containing the concave section of the pattern piece. This must be done before you can proceed with cuts 2 to 6.

2 The next cut after 1 is the score 2 along the edge of the concave curve. Then make a series of concentric scores parallel to, but outside, 2 (3 to 6). Tap along each of these in turn, moving inwards from 6 to 5, etc, and then gently separate it with the glass pliers. When the whole concave section has been removed then groze away any unevenness on the inner edge (line 2), making certain that it matches the contour of the pattern piece. (You can, of course, make more or fewer scores than indicated to remove the concave section. The more you make, the less pressure on the glass along line 2.)

3 Score and break off the glass along lines 7 and 8. Cuts 10 and 11 were not necessary in the piece of glass that I used because the width of the sheet of glass corresponded exactly to the length of the pattern piece. If the sheet of glass overlaps the pattern piece at either the top or bottom, then make cut 10 or 11, or both, at this stage.

4 Finally, score and tap line 9 to remove the excess glass to the left of the pattern piece. Take special care to tap line 9 so that it fractures with minimum pressure. The area between it and the centre of the concave curve (line 2) is where the glass is narrowest and thus most liable to break. It is for this reason that the concave cut was made before the straighter line 9. The more glass to support a concave cut the better: in this case all the excess glass to the left of the pattern piece reduces the likelihood of line 2 breaking incorrectly. If

The sequential cuts used for cutting out piece 4

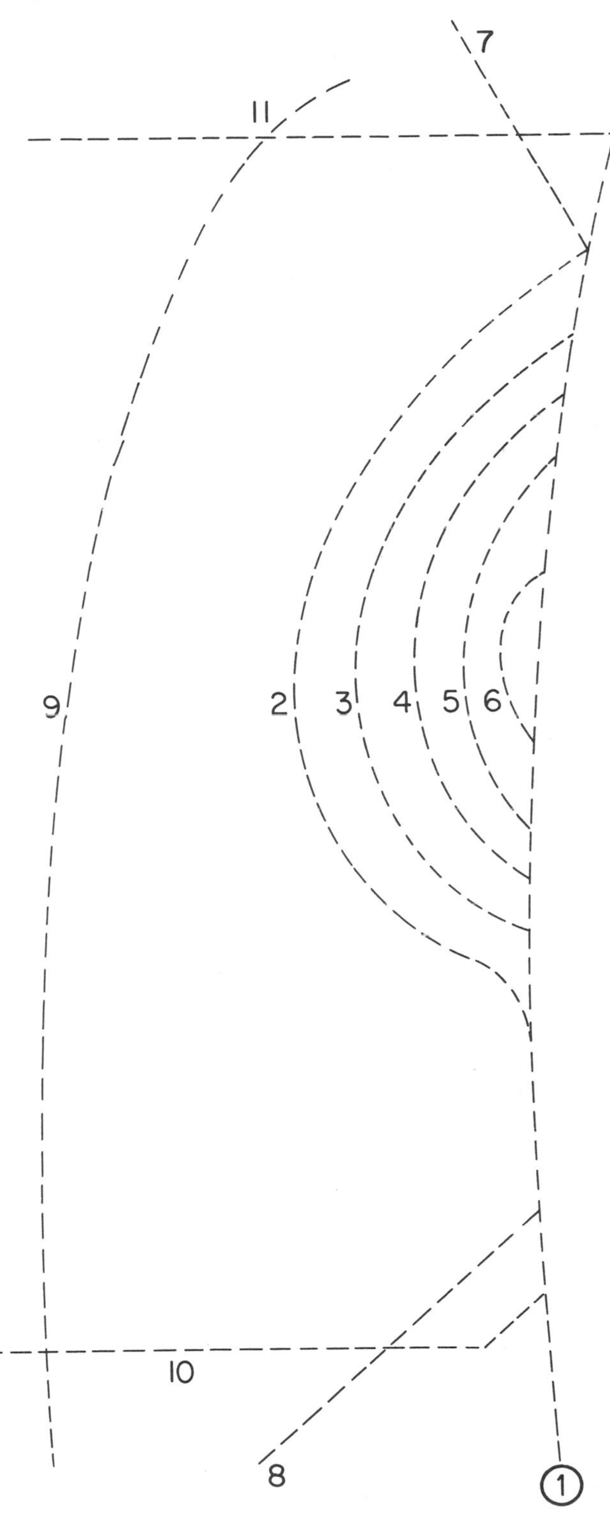

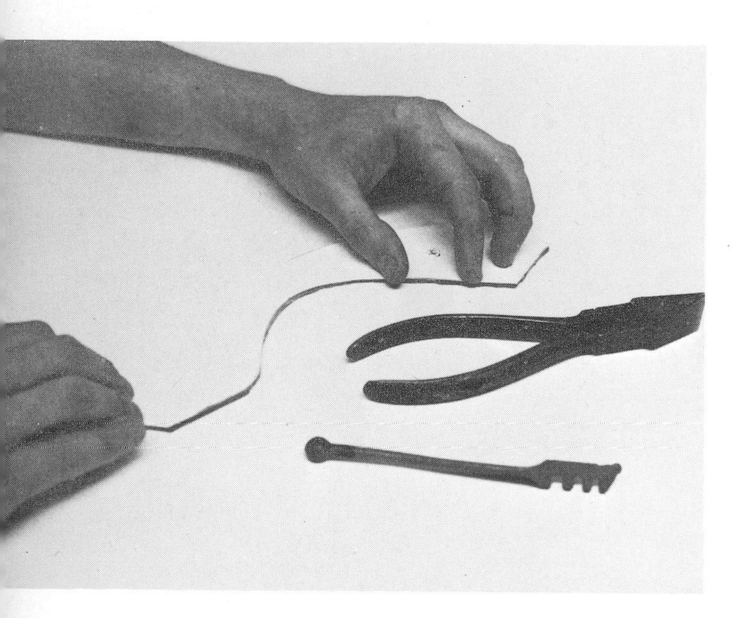

Make a final check to determine whether the piece of glass is exactly the same size as the pattern piece

line 9 had been scored and broken first, then the narrower area of glass giving structural support to the concave cut would greatly increase the stress on the curve. When you have completed line 9 then place the pattern piece on the glass to make a final check whether they are *exactly* the same size, grozing away any edges on the glass that overlap the pattern piece.

As you complete each of the seventeen pieces of glass then place it on its corresponding section on the cartoon, which should be placed on the work-bench alongside the area where the panel is to be assembled. You are at this stage actually pre-assembling the panel, but without the strips of lead. This is a good procedure as it will ensure that you have not forgotten or mislaid any of the pieces, which should, when they have been cut, exactly match the cartoon beneath them – allowing, of course, for the $\frac{1}{16}$ in. for the lead between each piece.

The next step is to assemble the glass in the panel. Place the work drawing on the work surface and frame it on two adjacent sides with two strips of wood so that the strips are flush against the outer perimeter lines, 15 in. × 10 in. on the work drawing. The two strips of wood must be longer than either edge of the panel (make the one about 18 in., the other 12 in.) and must be nailed through the work drawing into the work bench. Make sure, in doing this, that the strips are at right angles to each other. (The work drawing can, alternatively, be taped to the work bench. Whichever way it is secured, make certain that it cannot move, because it must not be allowed to shift while you are working on top of it.)

Next, take the $\frac{1}{2}$ in. H-lead came. Stretch it and then run the lathekin along both channels to remove any remaining kinks. Now cut two strips from the $\frac{1}{2}$ in. lead; one of length about 16 in. the other about 11 in. with the lead knife and place them along the strips inside the frame. They meet in the inner corner of the panel and can either butt up against each other so that the end of the one is at right angles to the side of the other, or they can both be mitred at a 45° angle. Either method, when soldered, will ensure a strongly bonded joint.

General guidelines when constructing the panel

The following are general guidelines that should be applied when constructing a panel:

1 Always assemble the glass so that you work from the inside corner of the wooden frame outwards in concentric circles. The first piece of glass fits into the corner and subsequent pieces border on it in ever-widening arcs. Think of the first piece as the cornerstone or apex of a pyramid that you are building. The second piece to be fitted should support the first by locking it into the perimeter leads and using it as the foundation for its own support, while the third will probably border on both the first and second pieces. And so on, piece by piece.

Try to determine in advance the optimum building pattern, with each new piece fitted being supported by the preceding pieces places in the panel. This ensures that whatever pressure is applied by tapping both the pieces of glass and the lead came so that they interlock firmly, will be properly supported. If this is not done, and gaps are left to be filled at a later stage, then the panel will, on completion, be structurally unsound.

Bear in mind in this respect, that leaded glass is not only an art-medium, but also an architectural one. The two are inextricably bound together: a beautifully designed, but badly-made window will be a short-lived pleasure. Some mediaeval windows have already survived a millennium, and with the modern materials and equipment available, work of today should be able to withstand the ravages of time for a considerably longer period. So build for posterity: every finished leaded glass object should embody the qualities of both durability and beauty. Glass is functional art.

2 Before starting the panel (or any glass project) make certain that the edge of the glass to be used is narrower than the height of the channels in the lead came. Glass that is too wide for a channel cannot be forced into it; the leaf on the lead will tear or be crushed inwards. This can be a problem especially when using antique, hand-blown glass as the edge can vary considerably in width along the length of a single sheet of glass. Examine carefully the edge of each piece to determine its widest point. Only then can you select the lead came to be used. If you neglect to do this before you start the panel you may find later that a piece of glass (invariably, one of the last to be assembled) will not fit into a channel. And no amount of cajoling or manipulation of the leaves of the lead with the stopping knife will overcome this problem. You will either have to

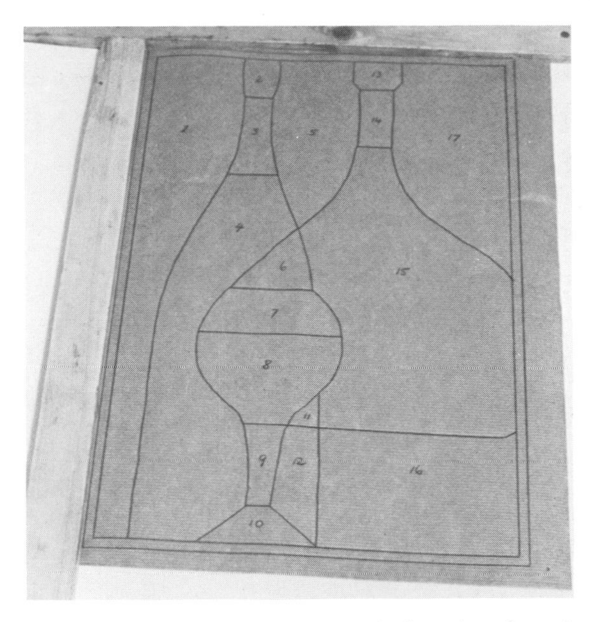

The work drawing framed on two sides by strips of wood that have been nailed to the work bench

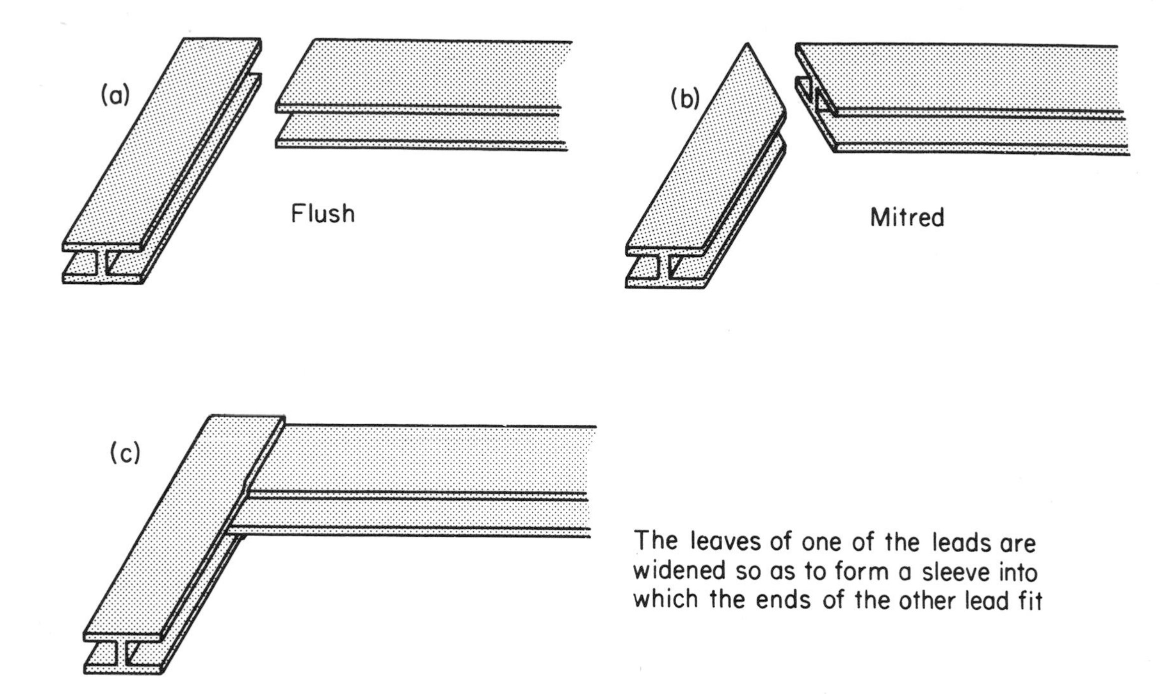

Three methods of joining perimeter leads

use another, thinner piece of glass (probably of another colour, which will, therefore disrupt the intended overall colour balance in the panel), or replace the lead with another piece that has a larger channel. Neither solution is satisfactory, so prevent such difficulties by checking beforehand.

3 It is advisable, especially in one's first few panels, to number the pieces on the cartoon in the sequence in which they will be assembled in the panel. Piece 1 is, therefore, the one fitting into the inner corner of the wooden frame. Sequential numbering helps you to work out in advance the step-by-step method of building up the panel.

A quick checklist of tools and materials is in order here so that you have everything that you may require to complete the panel: lead knife, stopping knife, grozing pliers, lathekin, nails (horseshoe, preferably), hammer, flux, flux brush, wire brush (or steel wool), soldering iron, and solder.

Now to start. Place the first piece of glass (1) into the panel so that it fits snugly into the channels of both the side and top perimeter $\frac{1}{2}$ in. H-leads. If necessary, apply pressure by tapping the glass with the back of the stopping or lead knife to ensure that it is firmly in place. If you feel that the glass is catching on the lower leaf of the lead and is not, therefore, inside the channel itself, then insert the blade of the lead knife under the glass to lift it over the lower leaf of the lead. When the piece of glass is correctly positioned it will line up along the line on the work drawing beneath it that corresponds to its inner edge. When you are satisfied that it does, then hold it in place with two or three nails. Next, stretch the two strips of $\frac{1}{4}$ in. round H-lead came, straighten out any bends in them with the lathekin, and then remove the nails and place the lead along the inner edge of the piece of glass. Cut the lead off $\frac{1}{4}$ in. from the end of the glass (to allow for the leaf of the perimeter lead at the bottom of the panel), and then secure it with nails. Use the stopping knife to press the heart of the lead flush against the glass.

Each strip of lead must first be measured, then marked, and finally cut, to ensure that it is the correct length for the space that it is to fill in the panel. Neither can it be too long – in which case it will get in the way of any leads against which it abutts; nor can it be too short – as this will leave a gap that will have to be bridged later with solder. You will, with practice, soon learn what is, and

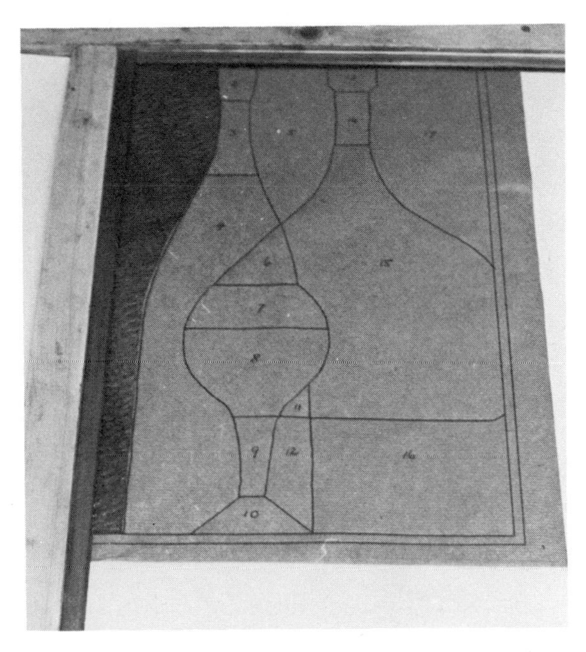

The first piece of glass positioned in the channels of the perimeter leads. Notice, also, that the glass does not extend beyond the line on the work drawing

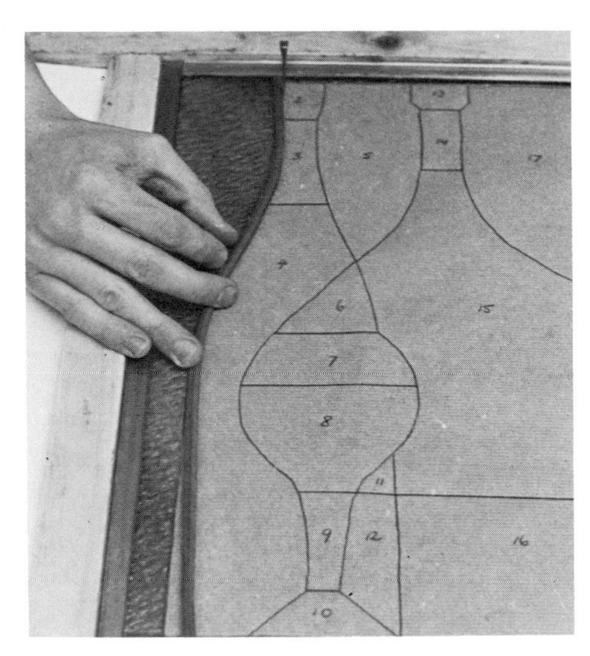

Inserting the first strip of lead came

what is not, the right length for a strip of lead. Not only should each piece be the correct length, but its ends should be mitred to ensure that they fit flush against the adjoining leads. Careful preparation at this juncture eliminates time-consuming adjustments and the likelihood of unsightly, messy joints when the soldering stage is reached.

The correct length of lead to use – whether to cut off each strip so that it holds only a single piece of glass, or to employ long strips that run the entire length or breadth of the panel or window – depends, of course, on your particular design. Bear in mind, however, that long, uninterrupted lead lines tend to be: (1) more monotonous and uninteresting than shorter or curved ones; (2) less supportive structurally; and (3) harder to work with the longer they are.

This can only be a generalization, though, as, in the right artistic context, a long lead gives a sweeping, rhythmic effect. In addition, in a small panel, such as in this chapter, too many leads give a visually cluttered effect.

Now proceed to the next piece of glass, piece 2. Position it firmly in the top perimeter lead and cut a piece of $\frac{1}{4}$ in. lead to fit between piece 2 and 3, the

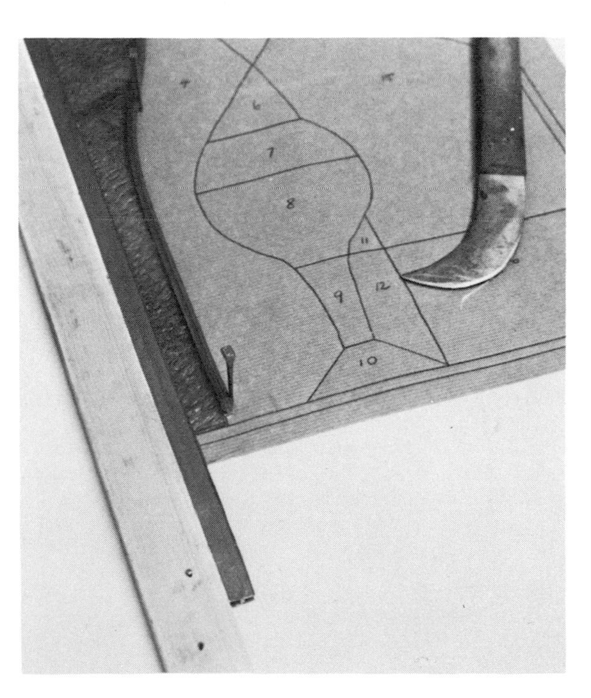

The first piece is cut off $\frac{1}{4}$ in. from the bottom of the glass and secured with nails

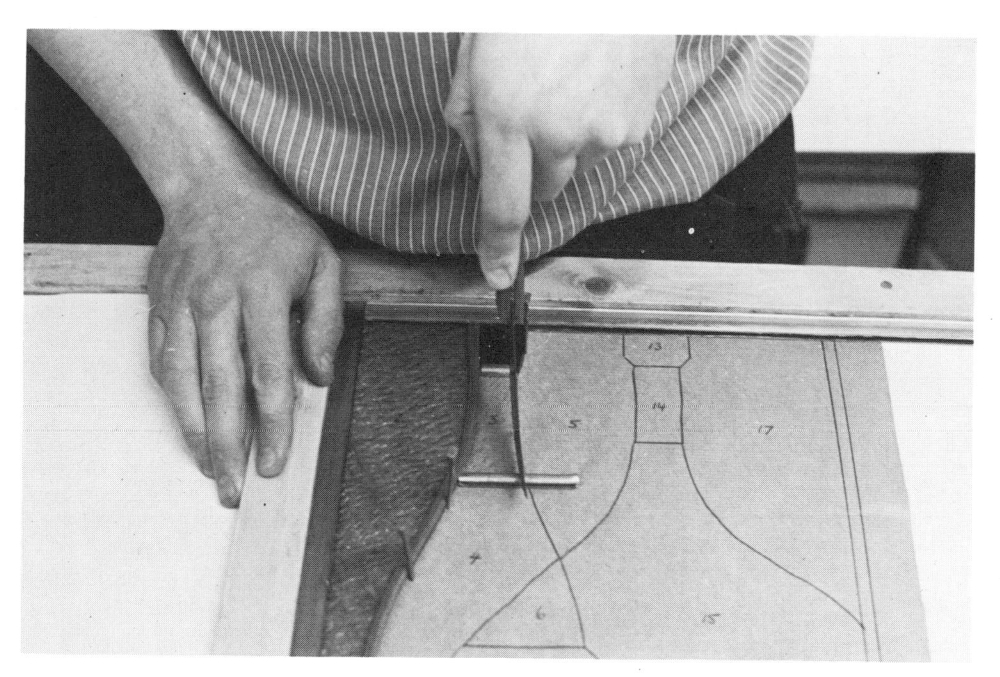

Marking the lead that will separate piece 3 from piece 4. The cut will be at a slight angle to allow for the curve of the line in the design

Using the stopping knife to push the lead up firmly against the glass

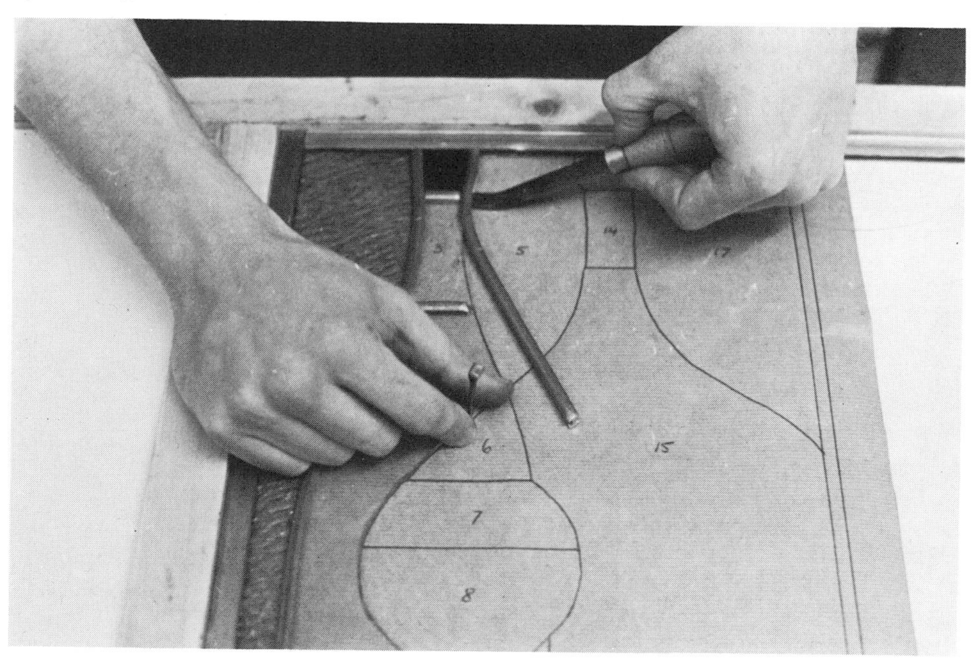

next piece to be put in place. The one end of the lead should butt up against the first strip of lead put into the panel, and, at the other end, be cut off $\frac{1}{16}$ in. from the end of piece 2. This is to allow for the overlap of the leaf on the vertical lead between pieces 2 and 5, and this $\frac{1}{16}$ in. must be allowed for in all subsequent lead measurements within the panel where $\frac{1}{4}$ in. leads abutt each other.

Use the stopping knife again, and then take the next piece, 3. Tap it so that it fits into the lead channels, and then mark and cut the next piece of lead to fit between pieces 3 and 4, remembering that the lead must be $\frac{1}{16}$ in. shorter than the length of the glass. Then use the stopping knife to press it against the glass, and secure it with a nail. The work drawing beneath the panel is a guarantee that the panel is retaining its predetermined dimensions as it is constructed piece by piece. Check continually that the heart of the lead is over the line marked for it on the work drawing. If not, check if the leaves on the lead have become buckled, thereby preventing the edge of the glass from fitting correctly into the channel. Remedy this, where necessary, with the lathekin or stopping knife.

Position piece 4 and hold it with nails while you cut the vertical lead that fits between pieces 2, 3 and the top of piece 4 on the one side, and piece 5 on the other. Then put piece 5 in its place and run a strip of lead along the edge of 5 and 4 for the entire length of the panel, and, when you are satisfied that it fits correctly, cut it and then secure it with nails.

Proceed in this way – by putting the piece of glass in position and securing it with nails; cutting the lead to fit between the pieces; positioning the lead and securing it with nails; etc – till pieces 6 to 10 (in this order) have been fitted into the panel. Remember that the ends of the leads must be cut off $\frac{1}{4}$ in. from the bottom of the panel to allow for the leaf of the bottom $\frac{1}{2}$ in. H perimeter lead.

Next, cut and fit a piece of lead to border on pieces 6, 7, 8, 9 and 10, and secure it with nails. Make certain that the top end of the lead is mitred so that it fits snugly against the lead against which it abutts. Donotforget the nails afterwards.

Next, put in pieces 11 and 12, plus the lead between, and the lead bordering on, both of them.

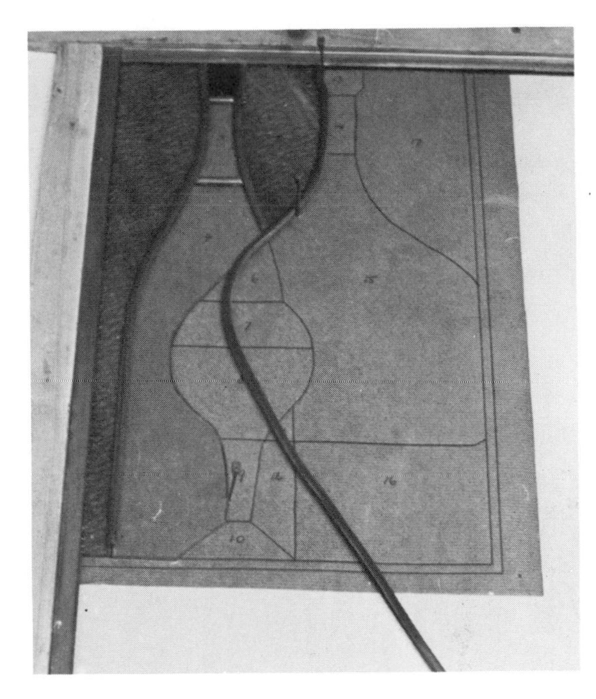

Secure the lead at the top of the panel with a nail and then proceed to fit it snugly against the glass for the length of pieces 5 and 4

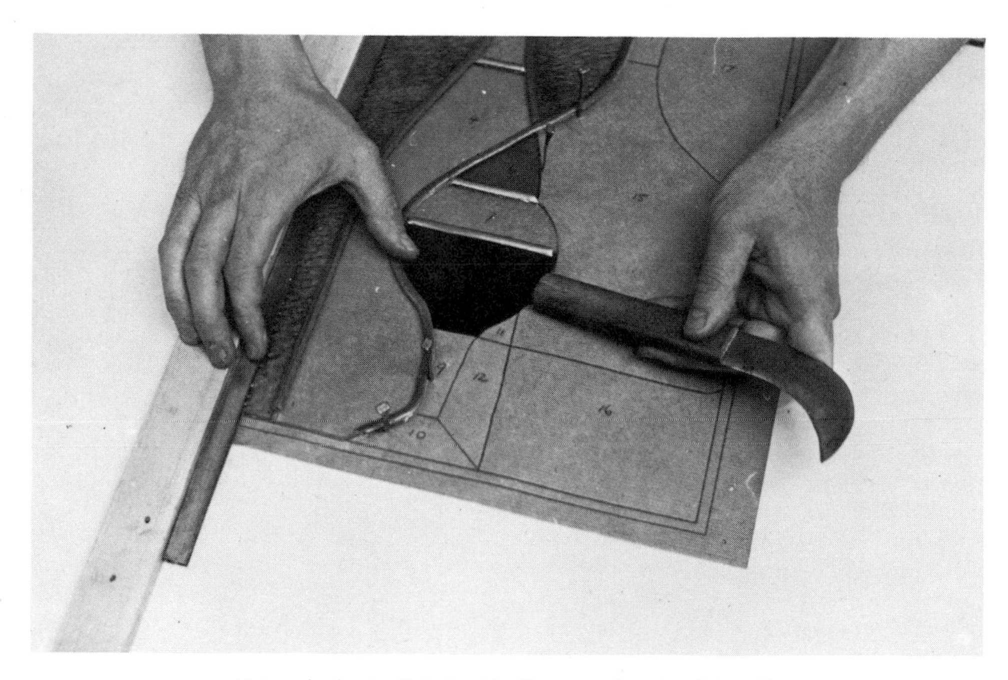

Using the back of the lead knife to tap the glass (piece 8) so that it fits firmly into the lead channels

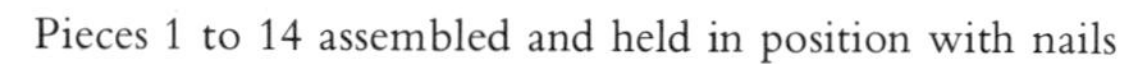

Pieces 1 to 14 assembled and held in position with nails

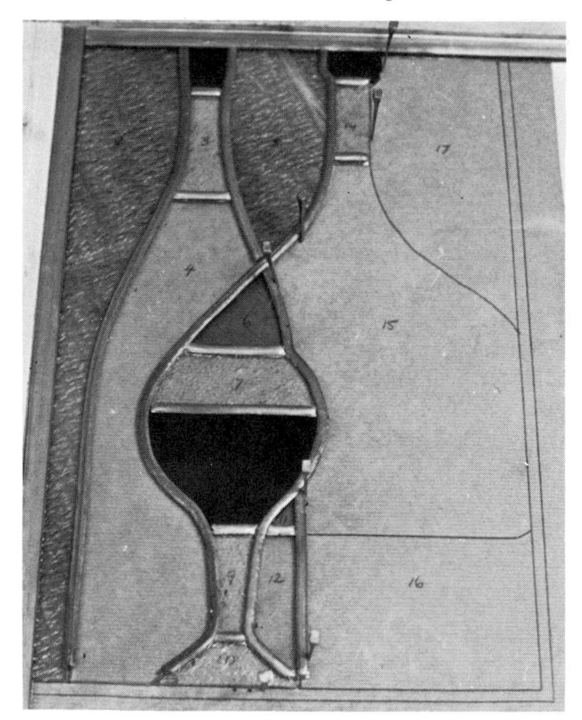

Now return to the top of the panel and, in the same way, put in pieces 13, 14 and 15. Make certain that 15 fits correctly around pieces 7 and 8. Check that all the lead channels are open before fitting it, and then, when it is in position, tap it with the handle of the lead knife to take up any excess space between it and the lead. Then cut and position the lead that fits between 15 and 16, and put in piece 16.

The panel is now nearly complete. Cut a piece of lead to separate 17 from 13, 14 and 15, put it in position, and insert 17. The inside of the panel is now completely leaded.

Make certain that none of the pieces of glass extend beyond the inner perimeter line $14\frac{1}{2}$ in. × $9\frac{1}{2}$ in. on the work drawing. The outer border of the glass should coincide exactly with the line.

Now put two outside perimeter leads on to the remaining two edges of the panel. First cut the one for the bottom edge so that it extends an inch or so beyond the edge of the panel. Put it in position and run the stopping knife along its outside channel to push the lead up tightly against the glass and secure it with nails. Then fit the remaining perimeter lead – again using the stopping knife – and secure with nails.

Finally, take the lead knife and square up the panel by cutting off the ends of the perimeter leads that extend beyond the panel.

The panel is ready to be soldered. Now is the time, ie, prior to soldering, if you are going to place the panel into a window space or some other set position, to make a final check that the panel's dimensions do, in fact, coincide with those of the predetermined measurements. In the event that the panel is too big then, if the adjustment in size is not too great, the empty, outside channel(s) of the perimeter H-frame leads can be shaved off to reach the required dimensions. Where, however, the panel is too small, then it must be extended. Neither should occur, though, if you made certain, during the assembly of the panel, that each piece of glass and lead matched its position on the work drawing exactly.

All but piece 17 assembled

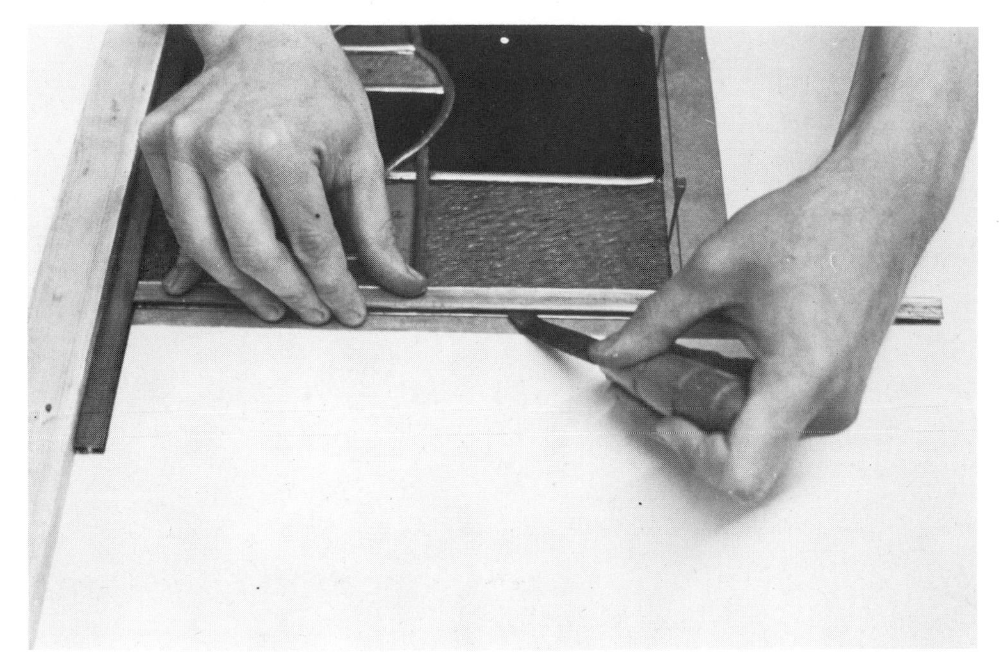

Pushing the lower border lead up against the glass with a stopping knife

The panel is almost complete, with only the border leads to be squared off and soldering to be done

The completed panel

Soldering the panel

This should be done as described in chapter 4. The sequence is as follows:

1 Make certain that the joints in the leading fit together snugly and neatly so that there are no large gaps requiring excessive solder.

2 Clean the joints with a wire brush or steel wool to remove the oxidization on the lead.

3 Flux each joint with the flux brush.

4 Before soldering the panel test whether the soldering iron is sufficiently hot. Put a piece of solder against the tip of the iron to see whether it melts readily. If not, check that it is plugged in correctly. Wait till it has absorbed more heat, and test it again. When it is hot enough then begin soldering at the corncr furthest from you (this prevents hot solder from dropping accidentally on to already soldered joints).

Progress systematically across and down the panel till all the joints have been soldered. Hold the end of the stick of solder right on the surface of the joint (ie, not suspended above it) and touch it with the tip of the iron just sufficiently so that a small blob of solder melts and bonds the joint. Be careful not to hold the iron on the joint for too long as it may cause the lead came itself to melt.

5 When *all* (check that none have been missed) the joints have been soldered, then turn the panel over and repeat the entire process on the other side. You will be able to determine on the reverse side just how accurately you are cutting and mitring the lead endings. Leads that appear to have been cut correctly on the front of the panel often do not fit as snugly on the reverse side because they have not been mitred exactly perpendicularly, leaving an overhang which results in larger gaps occurring between the joints on the reverse side.

The method of turning over a panel that has only been soldered on one side, thus minimizing the likelihood of its coming apart, is shown in figures A, B and C.

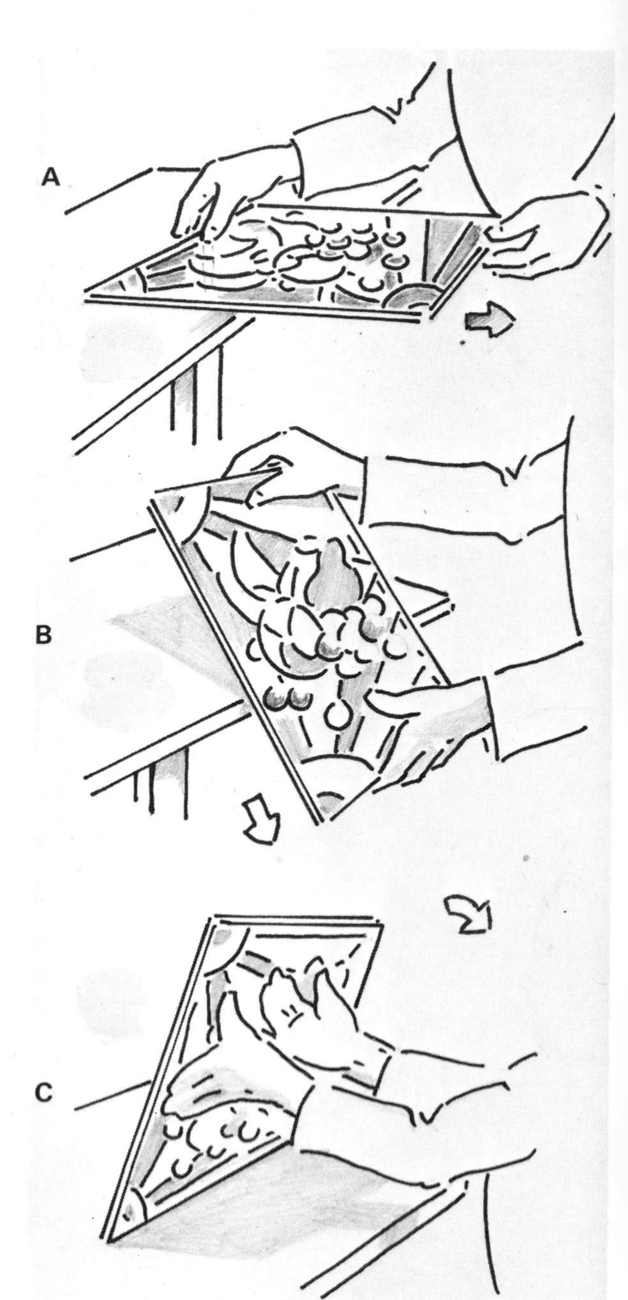

It may seem, as you prepare to turn your panel over, that it is virtually complete – requiring, if at all, only a perfunctory light soldering of the joints on the back of it. This is far from so, and many a well-executed project has disintegrated at this penultimate stage due to its being turned over too rapidly. So be patient.

The panel is now complete. If it is to be used as an outside window then it needs to be finished with putty, ie, weatherproofed, so that all the joints are sealed to withstand the elements. This is described in chapter 8. If, however, the panel is to be used indoors, then all that needs to be done is to wash off all the excess flux, solder and grit from the surface of the glass and joints with water and cleaning powder or liquid, and, when this has been done on both sides, to run the blunt edge of the stopping knife or some other instrument, along the edges of the lead came to ensure that the leaves grip the glass securely. Mention was made at the beginning of the chapter of the fact that several methods of displaying the panel exist. Here are two different uses to which it could be put, though there are, of course, many more:

Solder several panels together – again, varying the design and colours – to form a room divider that could be used, for example, to separate a small bar counter from the rest of a lounge

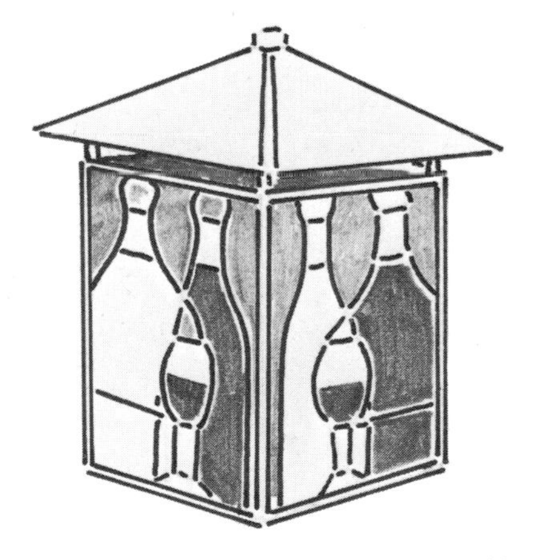

Assemble four panels of the same size. I would suggest varying the design and colour scheme on each one and soldering them together to form a lantern

Chapter 7

Design

It is important initially to realize that designing for glass is unlike designing for any other medium: glass, like everything else, has its own idiosyncrasies, and these must be taken into account before translating the intended design from sketch form into glass. All artistic media such as oils, tempera, gouache and charcoal do, after all, have certain intrinsic characteristics with which the artist must become acquainted; with the ones pertaining to glass being no more limiting than those in other art forms.

The principal limiting factor is a technique one: that certain shapes simply cannot be cut out of glass. Acute angles and long, thin strips are, for example, often impossible to achieve and are constructionally weak. So, where possible, these should be avoided at the design stage. By recognizing what can, and cannot, be done beforehand you can largely pre-empt and circumvent any such potential constraints.

If coming to leaded glass with a background in other art forms you are well advised to suppress such knowledge. It is not easy, for example, to make the transition from canvas to glass successfully (notable exceptions, however, being Leger, and Matisse), as a painter tends to rely on brush strokes to obtain his detail, whereas the leaded glass exponent utilizes the leading within his design to help give him the effect that he desires. A painter often, in fact, regards leading as a solely functional aspect of glass – as a necessary evil, almost – that he has to incorporate in his glass project in order to bond the pieces of glass together. Canvas is not, though, a complementary art form to glass, and you cannot move from the former to the latter without making an adjustment for the change in media (unless, perhaps, you are a cubist in the Mondrian mould).

Just as a composer can arrange the notes on the stave of his music sheet to produce innumerable different tunes, so in the same way a leaded glass artist has his building blocks with which to obtain his own preferred symmetrical, linear, and colour balance: namely, space, leading, and glass. Let us examine each of these in turn to determine how it can be employed within the design to maximum advantage.

Space within a design

A leaded glass artist must first determine the dimensions of the project that he is going to execute in glass. Once he has established these measurements – they could, for example, be the length and breadth of a window space into which he is going to place the window that he is designing – he must then plan how best to utilize this space by dividing it up at the drawing board stage.

There is no set formula for designing in leaded glass, though one is bound, on the one hand, by the limits of the materials and, on the other, by the extent of one's own creativity. The two are interrelated, as a good design is always an expression of the materials being used. It is, of course, impossible to put down on paper what is good and what is bad design: the viewer will decide for himself whether he finds the total effect aesthetically pleasing or not. All that can be offered are certain fundamental aspects of design that are, I feel, beneficial to someone new to the craft.

Spatial division and equilibrium are achieved in leaded glass by lines. How successful the end result is depends on whether the sum of the individual parts, into which the total space is divided by the lines in the design, creates a harmonious and interesting whole.

It is lines, therefore, that are of paramount importance to a leaded glass exponent as they divide up the total space and define the contours of the shapes of the elements within this space, while all the time directing the eye to and from each area in a systematic manner. Whereas the various tonal and textural characteristics of the glass will supplement this, it is primarily the way in which lines are used that determines whether the desired effect is, or is not, achieved.

A line's dimensions consist of its direction, length, and width. The last two, length and width, are discussed in the next section, and we will concentrate here on a line's direction.

The function of a line is, quite simply, to direct and control the eye as it examines the composition: it should lead to and from the focal points to encompass all aspects of the design, and the manner in which they interrelate. A line can be only one of two possible shapes: straight or curved. A straight one, due to its uniformity, imparts a sense of formality and monotony but it can, in conjunction with other lines – both straight or curved – be incorporated to create any number of eye-catching geometric designs. It is only on its own that it lacks inspiration.

Nuit de Noel by Henri Matisse, commissioned by *Time* Inc, 1952. The panel is charming and shows Matisse's complete understanding of glass as an art medium
Collection, The Museum of Modern Art, New York

Curved lines, on the other hand, are both beautiful and kinetic. They orchestrate the pace and movement of the design by the frequency and degree to which they change direction. A long rhythmic sweep, for example, conveys a sense of fluidity and gradual and predictable change, while a line that is made up of a series of short twisting curves generates a feeling of rapidity and unrest. A line does, therefore, have a preferred shape for each desired effect, with certain linear movements expressing certain sensory feelings. The direction in which lines move has a good deal to do with the expressive values that the viewer assigns to the shapes that those lines contain, and in this exercise an influence over where the viewer will look next. Partly by practice, and partly by studying this aspect of design in other people's work, you will soon be able to determine what impression a particular shape of line is likely to have on the viewer.

Curves, although being inherently more creative than straight lines, do themselves vary in terms of impact. The arc of a circle is the least interesting curve as it never changes direction, while a section of the outline of an ellipse is, for example, more beautiful than that of the arc as its origin is less apparent, being traced from two centres. The curve that follows the shape of an egg is, in its turn, more subtle than that of an elliptical curve since it is traced from three centres. And so on; with less obvious curves imparting increased sophistication to the design. Try always to avoid everyday shapes such as the perfect circle, triangle, or square as the eye has been conditioned to them and will not respond in as animated a fashion as it will to a more intriguing shape.

It is always a good idea if you are new to design and are working with standard shapes, to consider modifying them, however slightly, so that they have a measure of eccentricity, and hence unpredictability. A rhomboid is, for example, more interesting than a square, and an irregular circle likewise attracts the eye more than a perfect one. Drawing shapes freehand will often, in fact, create such 'imperfections' which, in turn, give your work some individuality, and this is the first step on the way to what is, I feel, the quintessence of art: to create designs which are an expression of your own, and only your own, personality. So, in those instances where you have tried to reproduce an existing design, but were unable to do so accurately, console yourself that whatever difference exists between what you were copying and what you produced, is your own creation. To pursue this in general terms, the more originality in your work the more creative and therefore artistic it will become. It is, I feel, the presence or absence of individuality in one's work – as discussed in the Foreword to the book – that determines finally whether someone working in glass can be described as a leaded glass artist or craftsman. The difference is that simple . . . and that hard.

It is important firstly to determine at the design stage what and where the composition's focal point will be, so that each line included in the design will have a specific function relating to this point. There are two things to remember in this respect: first, in order to emphasize a single aspect of the design, it is important to ensure that the other parts are less pronounced. The background's primary function is to lead the viewer, by the judicious use of the lines in the composition, to the focal point. So make it less accentuated, both in terms of the number of lines and in the colours of the glass, than those at the central point. Too much background detail can confuse and exhaust the eye by presenting it with a mass of unco-ordinated lines which appear to run all over the glass in helter-skelter fashion. So save your intricacies in line detail and most striking colours for the main parts of the design.

Secondly, and partly related to this last point, is that while planning the lines in the design, constantly remind yourself of the limits of the medium. Try to keep each area of colour within the limits of what can be cut out of a single piece of glass. It is, in fact, a good general rule to make every lead in the design the excuse for a change of colour or tone, however subtle. Lines that run through a monochrome tend to both detract from the intended artistic effect and also to highlight the structural limitations of working with glass. It shows, for example, that you could not cut acute angles; that whereas you would not by choice break up a monochromatic colour that the inherent properties of glass necessitates your doing so. So soften any sharp, potentially uncuttable corners in the lines before committing the design to glass. This will, in addition, often add fluidity to the overall movement of expression by ironing out any unintended jaggedness.

An old panel, in a San Francisco bar, containing only two different colours of glass; the charm of the design lying largely in the effective combination of straight and curved lines

Certain shapes convey certain sensory feelings to the viewer. Diagram adapted from *The Advanced Techniques of Making Leaded Glass Projects*, by Walter J Hamilton
Reproduced by courtesy of the author

A magnificent example of the effective use of straight lines used in conjunction with curvilinear ones. The total visual effect is one of continually evolving fluidity of movement. Notice, in addition, the virtuosity of the leading and the exactness of the glass-cutting. Detail from the west wall, St Johann, Dortmund-Kurl, Germany
Ludwig Schaffrath, 1972

A modern window showing the effective use of curved lines to impart a feeling of continuity of movement. The eye is propelled from the bottom of the composition upwards by the flowing motion of the lines in the design
Alfred Fisher, Kings Langley, Hertfordshire

Leading within a design

Leads are the material means of translating the lines in the design into glass. They do, in addition, however, provide the sophisticated glass exponent with a means, by the interchange of lead widths within his composition, of extending the range of artistic effects that he can achieve. In fact, varying lead widths allows one to implement so many subtleties into one's design that it constitutes one of the most esoteric aspects of the craft. Always bear in mind, therefore, that leading has a dual function: that of containing the elements of the design *and* of doing so artistically. The former aspect was discussed in chapter 4, and we will now look at the latter to see how best you can apply it within your work.

First, develop the habit, on completing the sketch of the intended design, of drawing in the lead lines so as to give an accurate representation of the overall finished project. By this I mean that where, for example, you intend to use a $\frac{1}{4}$ in. or $\frac{5}{8}$ in. lead in the design, that you first draw it *to scale* on the sketch (which will, in turn, be transferred on to the cartoon) to give as exact a representation of the finished panel as possible. (If the sketch is, say, a half the size of the final dimensions of the panel, then you should, of course, draw the lead lines to a half of their actual width.) It is only in this way that it will be possible to gauge what effect your choice of leading will have on the total balance of lead and glass in the composition, and it is the only stage at which you will be able to experiment effectively with different lead widths to determine which is the optimum lead size for each area of glass in the design. To interchange leads while assembling the panel is not only invariably haphazard and unsatisfactory from an artistic and technical viewpoint, but also results in the quite unnecessary cutting up and eventual wastage of expensive lead cames.

It is a popular misconception amongst both the public and most glass exponents themselves that certain detail – such as that required for facial features – simply cannot be achieved in glass without an application of paint. This, as the photograph so convincingly shows, is untrue.
Portraits of Bob Dylan and Mick Jagger in leaded, unpainted glass *by Jeff Speeth, Friendship, New York*

What the newcomer to leaded glass cannot always immediately appreciate is that, due to his inexperience, leading can take up considerably more or less room in his panel than he had initially intended. It is only with experience – one tends, in this case, to learn mostly from one's mistakes – that one can visualize the ratio between the leading and the intervening space in a composition without actually having seen it in advance. Small pieces of glass, in particular, are likely to be swallowed up by the bordering leads unless allowance is made for this by reducing the lead size accordingly. Conversely, large pieces of glass, if bordered by thin leading, will cause the linear detail that the leading affords to be minimal. If the design calls for a spidery effect between large pieces of glass, so well and good, but if not, then a correspondingly larger lead size should be incorporated into the design at this point to balance the optical illusion caused by the larger areas of space. It is a good general rule that the smaller the pieces of glass, the thinner the leads; and the larger the pieces, the thicker the leads, but, once again, any number of spectacular effects can be obtained by intentionally disregarding this observation.

If, as a beginner, you simply draw in the lines on the sketch without considering at this stage what combination of lead widths best suits the design, you will probably soon forget that leading constitutes a very important variable in the overall design potential of your work in glass, and this, in turn, will lead to another habit that militates against creativity; namely, that the home craftsman frequently goes to his lead supplier to purchase a certain number of strips of lead all of the *same* width, which he then uses until he runs out, at which stage he goes back to the supplier and reorders another batch. This is to use lead at its lowest level; as a totally functional material.

Avoid this artistic pitfall by determining at the sketch stage what lead sizes, and how much of each, will be required for each specific project. The following are general rules that apply to the mixing of leads. Remember however, that these rules are, on the right occasion, there to be broken.

A design for a restaurant window, showing the breakdown of the total space into cuttable pieces of glass

1 Use very thick leads ($\frac{1}{2}$ in. and up) with the utmost caution and discretion in small panels and windows as their over-use will overpower the thinner leads. In the same way that a very loud instrument, such as a gong, should be used minimally in a musical score for maximum impact – as if it is used too much it will drown out all the other instruments in the orchestra and also lose its intended strident crescendo effect – so too with thick leads; if used injudiciously and too frequently they detract from the delicacy of the thinner leads and leave the viewer with a feeling of being visually battered (although the effect may be subliminal in that he or she may not know exactly why).

2 Develop the habit of varying the lead widths wherever possible to increase the detail in the composition. In the same way that a painter varies the width of his brush stroke to distinguish between major and minor lines in his painting, so a leaded glass artist should vary his lead widths to achieve the same effect. A panel that uses a uniform lead width throughout tends to have an uninspired, sluggish appearance; one which is unlikely to attract or hold the eye's attention for long. Therefore, to avoid such monotony you should incorporate different leads – $\frac{3}{16}$ in., $\frac{1}{4}$ in., $\frac{3}{8}$ in., etc – in the design so that they help the eye to identify which are the primary, and which are the secondary, lines in the design. This point cannot be overstressed as the use of a heavier, ie, wider, lead to emphasize the major movements in your composition, and of a lighter lead to indicate the subordinate ones, is the only way that leading can show the weight or balance that you apportion to each section of the linear aspect of your design.

3 Having just stressed how important it is to interchange lead widths, I must modify this by stating that it should be done with a sense of gradualism. Do not, for example, without careful consideration, run a $\frac{3}{16}$ in. lead into a $\frac{1}{2}$ in. one as the transition is normally too abrupt, with the result that it is jarring on the eye. Leads of different widths should be phased into each other gently so that the viewer's eye will comprehend the reason for the change in lead sizes without having to come to a complete halt. There is a finality about a very thin lead abutting a thick one – the effect is that of the thinner lead having come to the end of the line, so to speak – which, except in those instances when this is the desired effect, leaves the eye suspended in space. So mix the leads with the intention of varying the pace of the linear rhythm without bringing it to a stop. Avoid abrupt, discontinuous effects. In the same way that a music expert will detect a single discordant note in a concert piece, so an unbalanced lead width within what is otherwise a perfect glass composition will be equally evident to an art critic. What cacophony is to music, disharmony is to leaded glass. So remember, when running one size of lead into another: *pianissimo.*

On those occasions, however, when the design calls for an agitated or spasmodic effect to engender a sense of shock or dynamic action, then you should, of course, convey this by means of striking contrasts in your lead sizes. Make certain, though, that you get the right balance by experimenting at the sketch stage.

4 The length of the lines in the design determines, of course, how long each strip of lead will be. Remember, at the one extreme, that a long lead, if not touching on other leads, other than at its ends, does not supply much structural support and is, especially if it is straight, often less interesting than shorter, undulating ones. Very short leads, on the other hand, whereas they are structurally sound, can give a cluttered, unplanned appearance and greatly reduce the amount of glass that is visible, thereby distorting the intended artistic effect.

5 Light that is transmitted through glass can cause the width of the leads in your project to appear thinner than they actually are. This visual distortion is known as halation, which can be defined as the spreading of light beyond its proper boundary. Halation occurs when the intensity of the light is particularly fierce, such as solar rays striking the glass directly, with the result that the light diffuses beyond the edge of the glass, giving the intervening strips of lead came a hazy, less definable, appearance, and making them seem narrower that they in fact are.

In general, you need only concern yourself with this phenomenon of light if you are designing a window which will be positioned so that it catches the sun's full rays. In this case you should make allowance for the halation factor by slightly increasing the lead widths in your design.

An example of variation in lead sizes. The central linear movements are emphasized by thicker leads, while the less pronounced detail is shown by thinner ones. Note, also, the neatly soldered joints. Detail from the east wall, St Marien Catholic Church, Bad Zwischenahn, Germany
Ludwig Schaffrath, 1970–71

Designing for steel core leads
A steel core lead (described in chapter 4) is a lead came that contains a steel bar inserted into the heart of the lead to reinforce it. If the size of the project, such as a window or room partition, is greater than approximately three feet square, then steel core leads should be inserted every 24 in. or so to support the glass.

Cored leads create, if not problems, then certainly constraints, on a leaded glass artist's design potential, and it is imperative that, if the size of the project is such that it requires reinforcing, that this is anticipated and allowed for at the design stage. The reason for this is that, due to the structural strength that the steel imparts, so it has limited flexibility. It cannot, simply, be inserted at one side of the, say, window, and then be bent freely, as can a normal H-frame lead, to follow the contours of the leading in the design as it progresses across the window to the other side. Steel cored leads do have a certain degree of pliability, but maximum strength is provided when the bar is straight or only slightly curvilinear. They can be incorporated into the design either horizontally or vertically – the choice depending largely on the project's dimensions. If it is a standard rectangular window then the cored leads are usually inserted horizontally, but if the window is wider than it is high, then more strength may well be supplied by placing them vertically. So always analyse your design in terms of its size and whether it will require support *before* committing it to paper. If you find that you will need steel core leads, then you must start with these and build up the design from here onwards. Whereas you may well find that such discipline creates annoying design restrictions, this is far less frustrating than having to add reinforcement, which can totally disrupt the linear and colour balance, once the design has been finalized.

Steel core leads are not, incidentally, only necessary for projects that may be exposed to high wind forces, such as windows. Even indoor objects, for example, room dividers, need bracing to counter any jolts that they may encounter and to support the weight of the glass.

Glass within a design

Glass has two characteristics that relate to design: colour and texture.

Colour
Which colours of glass to choose to best suit the design is not something that can be solved by the application of a set of rules. With colour, more than with any other design factor, it is your *own* feeling for harmony that is the final guiding principle. Rules of colour mixing can insure against positively bad colour effects, but the end-result depends largely on your own particular colour perception, and is thus a subjective process.

The problem can be defined as that of trying to obtained variety in unity: of how best to achieve an interesting polychromatic effect within a balanced composition. It is not easy: a life-long learning process, in fact. Even for the most experienced professional colourists in the stained glass industry colour sensitivity is an ever-maturing process, one that they have to refine continually to meet the requirements of the design at hand and the texture of the glass that they are using. So do not despair. There is no such thing as *the* optimum colour balance. There are always, however pleasing the total effect, colour nuances that, no matter how subtle, could be just that bit better. It is, therefore, a matter of degree, with the key to betterment being to sharpen your colour sensitivity by whatever means you can: by continual practice and self-criticism; by appraisal of other people's work; and by careful analysis of colour relationships that you find particularly appealing.

What, then, can be done to help in this learning process? The best aid is the conventional colour theory, illustrated usually by means of a colour wheel. This gives the breakdown, by wavelengths, of that part of the total electromagnetic spectrum that is visible to the human eye. Some cautionary advice, though, on the use of a colour wheel for our purposes: first, it treats colour in the abstract and is for this reason often unsatisfactory as its concepts of colour harmony are not always applicable to the situation at hand. Secondly, it is less relevant to leaded glass than to other art forms, as colour theory, as defined in books on design, refers to the interrelationship of colours that are seen by reflected light rays, whereas in our case colour is viewed by transmitted light (how much depending, of course, on the amount of opacity in the glass). The glass is interposed between the physical

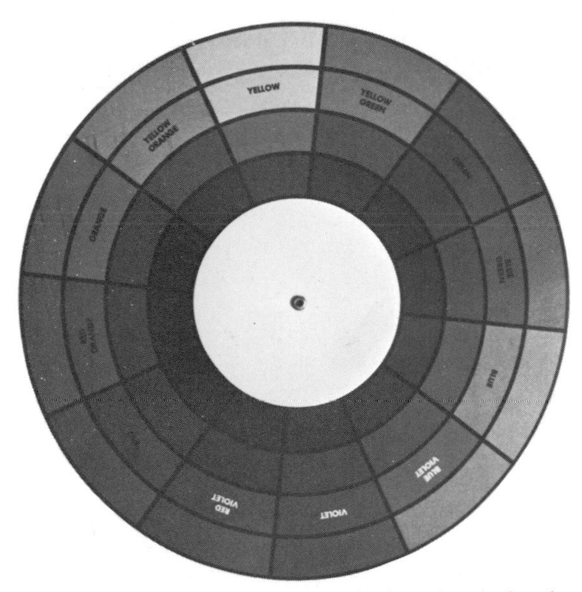

The bottom section of a colour wheel, giving the breakdown of the spectrum into twelve parts. The top section consists of an overlay that is positioned on top of this, and which, when rotated, shows in turn the analogous, complementary, and triadic colours for each of the twelve colours. The complete unit is called a colour harmony wheel and can be bought from art supply stores

energy of light and the eye so that it is not the surface of the work itself that is seen, but rather the interruption of, and modification to, direct light brought about by the colour and texture of the glass. The colour wheel is, notwithstanding, the best aid available, and when used in conjunction with a light table (below), gives a good approximation of the finished light effect, whether natural or artificial.

Colour theory, at its simplest, involves the division of the spectrum into twelve parts. These consist of the three primary colours; red, yellow, and blue; the three secondaries; orange, green, and violet; and the remaining six intervening colours, called *tertiaries*.

Each colour can, in turn, be further defined in terms of hue, saturation, and brightness. Briefly, hue is what we commonly mean by colour: for example, red, green, yellow, or intermediate shades; saturation is the relative purity of that colour; and brightness is its intensity.

It is interesting to examine certain characteristics of the above aspects of colour to determine what effect they have on each other when juxtaposed to form a multi-coloured whole.

Colours are rarely used on their own in leaded glass, but usually in combination, and one of the major problems in using different colours is that of their interrelationship. In order to know what a colour will look like in a completed project, it is necessary to know not only what its actual hue is, but also what effect the interplay of adjacent colours will have on this hue, as the contrast between colours in the same environment affects how they are perceived. This entails the mechanics of colour relativity, which involves the psychophysical aspect of light: namely, the relationship between the physical stimulus received by the retina of the eye and the subsequent experience reported to the brain. If, for example, pieces of red and blue glass are juxtaposed, then, whereas they are seen physiologically as red and blue, they are perceived psychologically as violet.

Another aspect of colour relativity is that of simultaneous contrast. The basic principle is that whenever two different tones come into direct contact, then the contrast will intensify the differences between them, with the change being greater or smaller in proportion to the degree of contrast. This means that two colours that are analogous (ie, similar) in hue, saturation, and brightness, will not have much effect on each other. Where, however, the contrast in these dimensions is strong, then the change in visual effect, in what is *perceived*, is proportionately greater than one would expect.

In most of your designs you will probably try to steer clear of the difficulties that the above can create by mixing complementary and analogous tones 'safely' to ensure that they blend. There may, though, be occasions when you will want to create the reverse effect: to mix strongly contrasting colours within a composition which must, nevertheless, be balanced in its entirety. This is much harder to achieve, and needs careful analysis before committing your final colour scheme to glass.

Certain colours simply will not go together. Where hue and brightness clash then the outcome is painful to the eye. Yet there are times when such an acid combination is exactly what you may want to convey. The problem is to control the contrast so that it does not tear the composition apart, but still retains all the force and vibrancy it can give. Two ways of achieving this are suggested: first, by size of area. One cannot generalize that large areas have stronger attraction than small ones. The attraction of a particular area depends not on its size, but on its qualities, such as colour, tone, shape, and position. Small, loud, stabs of colour can, therefore, be effectively employed within larger areas of more neutral tones without upsetting the overall balance.

The second method is by using contrasting colours in isolation. This, in fact, already occurs in leaded glass because each piece of glass is bordered by lead came or copper foil. Leading acts, due to its forming a partition between adjacent colours, as a neutralizer which helps to hold the contrast within manageable limits, with wide leads, of course, isolating the colours more than narrow ones. Fourteenth-century Gothic church windows are good examples of this isolation factor. The saturation of the red, blue, and green light would be too overpowering without the leading. With it, they combine to form a jewelled brilliance.

Although black and white constitute 'absence of colour', they are of great value to the glass artist. Here black, grey, and white opalescent glass are combined to form an unpainted portrait of Abraham Lincoln as part of a panel to celebrate the American bi-centennial in 1976. (The leads in the black areas of the glass cannot be seen but they are there)
Nervo Studios, Berkeley, California

Texture

The above section described colour theory for glass in terms of monochromes; eg, the effects of placing a piece of glass of a single colour next to another one of a single colour.

Most of the glass used by leaded glass artists, however, consists of many more qualities than simply one colour. The total effect of these is to make the glass infinitely more interesting than a plain monochrome by varying the effect of the light as it is transmitted through the glass. Exactly the same concepts of harmony apply, and you must still ensure that the colours balance. It is just that now there are more variables interacting to increase the radiancy and complexity of the overall effect.

Such properties of glass are its degree of translucency, its surface and/or internal texture, and its colour pattern. These were discussed from a structural viewpoint in chapter 2, but mention should be made here of their effect on design.

1 *Degree of translucency* Whether to use glass that is transparent or opaque, or that consists of some degree of translucency in between these two extremes, depends on its end-purpose. If, for example, you plan to design a window, then the decision whether to use clear or opalescent glass should include consideration of how much light the window is to transmit. If it is to be more decorative than functional, then you can use a more opaque glass than when maximum light is required.

Opinion differs amongst professional stained glass artists as to the degree of translucency that a window should have. Some feel that the eye should stop at the glass so that the spectator is not distracted by what is in the background (this has been termed the *Wall of Glass* theory). Others, however, feel that a window's primary function is to transmit light, and that any interruption or modification to this light should be done minimally. You will, with experience, be able to decide which you prefer, but try not to compromise by mixing opalescent and transparent glass within a single composition as this is likely to give a very disrupted, unplanned effect.

For lampshades, or any other three-dimensional object which has its light source inside the glass, I feel that the choice between the two is more obvious. The greatest degree of opacity should be chosen within the range of colours required. Lamps tend to be less attractive if the naked bulb is visible. The glass should intercept and trap the light rays so that they illuminate its innate properties without showing their own origin.

2 *Surface and/or internal texture* Some American machine-rolled glass is manufactured with various three-dimensional surface patterns, such as nodules, ripples, and ridges, while some English handmade antique glass has an undulating, reamy surface, or a mass of internal bubbles or 'seeds'.

Uneven surfaces not only provide glass with a tactile quality, but can also have an extraordinary refractive effect that disperses the light into myriads of particles, greatly enhancing the total appearance of the glass. Remember, therefore, when selecting glass, to look for such irregularities.

3 *Colour patterns* Some decorative glass, whether transparent or opalescent, has several colours impregnated into it during manufacture to give it a more interesting, multi-coloured effect. This admixture of coloured pigments is usually patterned in lines or bands of colours that swirl at random across the sheet of glass, giving the appearance of long, rhythmic brush strokes. The result, in transmitted light, can be breath-taking. Glass that contains such striations is known as *streaky*, and is the most popular glass for our purposes because of its variegated effect. One bit of advice, though, on its use. It is a seemingly trivial point, and one that to the unitiated eye would perhaps go unnoticed, but consistency in colour form is one of the factors that makes the difference aesthetically between an 'arts and crafts' object and a really fine piece of artisanship. To give an example: when making a lamp that contains a flower design and streaky glass has been chosen for the petals, then it is not enough merely to take a piece of that glass, cut the petals out of it, and then assemble them. They will probably all appear the same colour and seem congruous. Each piece should be cut to ensure that the striations in each petal give the finished flower a balanced appearance.

The effect for leaves and stems in a floral design is even more pronounced. Here the lines of colour in the glass must follow the natural movement of the leaf or stem to give fluidity. Nature will, incidentally, always give a lead in this, and if you are uncertain in which direction the striations should run, then examine a plant similar to the one you are designing in glass.

Light table

It will soon be realized that what you thought were the properties of the glass that you chose at your glass supplier can be radically different from those in the finished project. In the first instance you may well have examined each piece of glass that you bought individually and by reflected light, whereas the colour of each piece of glass in the final composition is influenced partly by the fact that it borders on other colours and by the fact that it is viewed by transmitted light. Failure to make allowance for such light modulating factors can totally destroy the intended colour harmony of the design.

In order to minimize such errors a light table should be used. In the same way that a painter mixes pigments on a palette to obtain his preferred colour balance before committing himself to canvas, so you should interchange pieces of glass over a light source to ensure that they are what you thought they are. Do not, though, in doing this, go to the extent of cutting each piece of glass into its respective shape in the design as this will entail unnecessary wastage if you decide to change it. Simply place small sections of the glass that you think you will need next to each other to get an approximation of what the end result will be. If the composition is to be illuminated by artificial lighting, then the light table provides a very good similation. If, however, natural lighting is to be used, then you must take this into consideration by holding several pieces of glass against a light source simultaneously. Alternatively, a more elaborate and professional method that you can use is that employed by several English stained glass studios. The pieces of glass are stuck on to a vertical sheet of plate or float glass with plasticine so that they can be viewed by daylight. This is an extremely effective way of checking your initial colour scheme for projects that will be seen by natural lighting as you are in fact pre-assembling the project (only the leads are missing).

If you do not have access to a light table then compromise by holding several pieces of glass against a light source simultaneously. Whichever method you employ, try not to rush your final decision as this is the last practical opportunity of achieving your desired colour scheme (to interchange glass while leading up the composition is invariably unsatisfactory).

When using transparent glass in a lampshade it is best to soften and diffuse the glare of a naked bulb by covering it with a globe (see chapter 12)
Lamp by Edward Russell, Digswell, Hertfordshire

Summary

This chapter has covered the components of leaded glass from a design viewpoint. Concern for design must, however, extend beyond the dimensions of a composition in glass to take account of the architectural and interior design environment in which the object will be situated. One of the most common mistakes made by amateur leaded glass craftsmen is that they create and then execute a beautiful design in glass without due consideration of whether it will blend into its final location. It is natural that when you first work in leaded glass that you experience a great sense of accomplishment in simply completing the project at hand. In the long run, though, this is not enough. Consideration must also be given to what is the project's end-purpose.

Whenever possible, determine in advance whether the intended choice of colour, linear style, and shape of the project matches that of the surrounding décor. A good professional stained glass artist will always insist on seeing the site of his commission before he begins his design, and so too should you.

This applies not only, most obviously, to windows, but to all glass forms, whether two- or three-dimensional. In the same way that a brightly contrasting abstract window will be totally incongruous in a Georgian styled sitting room, so too will an Art Nouveau floral lampshade when placed amongst modern furniture. In the first instance the window should portray a more traditional formal or pictorial design while in the latter the shape and style of the lampshade should be contemporary.

Leaded glass is such a new form of secular (as against religious) art that its effect on home and office interior design is largely unexplored. Very little literature is available on the psychology of modern architectural spaces, and until more is forthcoming I would recommend conservatism in your choice of colours. A churchgoer only spends an hour or two per week in a church, so an ecclesiastical window can be garishly bright and still not leave him exhausted. If, however, you have to work or live all day long with a window then it should be much more subtle. For example, a modern pastiche design comprised of gaudy colours will, in most cases, be too powerful for everyday living. Windows should, therefore, be very controlled in form and colour to avoid being visually abrasive. This is not, though, to suggest that you should forgo any combinations of colour schemes that you like, however loud or harsh; it is just that they should be used selectively, and not all be crammed into every window.

An example of colour inconsistency. Five panels of glass laid out prior to assembly into a lampshade. The direction of the grain on the middle panel runs horizontally, while those on the first and fifth panels run vertically. The middle one should be re-cut

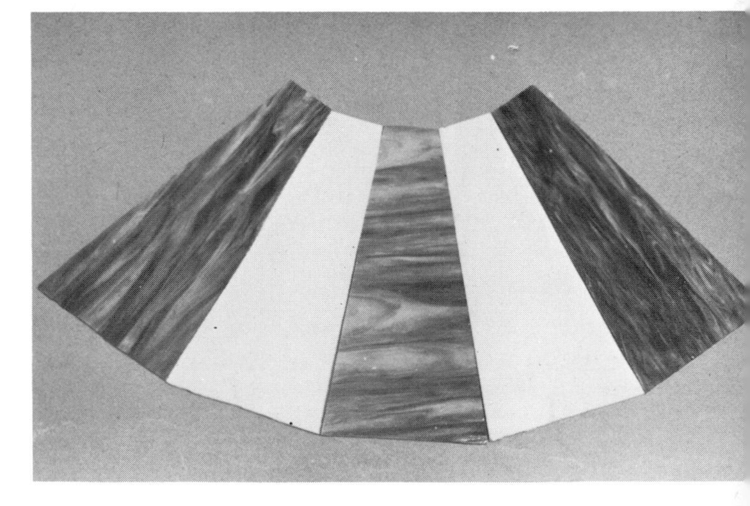

Chapter 8

Windows

Windows are, of course, in effect panels, such as the introductory one described in chapter 6, with the only distinction between the two being that people tend to think of windows as 'big' panels. The technique by which they are assembled is, therefore, the same as that in which the panel in chapter 6 was built up. But if, on completion of your first panel and after reading chapter 7 on Design, you are inspired to go on to 'bigger' things – not only in terms of size but, hopefully, in terms of artistry – then there are two additional aspects of panel- and window-making about which you need to know: namely, weatherproofing and reinforcing.

Weatherproofing

Windows are weatherproofed for two reasons: both to seal off the elements (which also stops the glass from rattling) and to supply structural rigidity. Do not be misled into thinking that a composition of lead and glass which is to be positioned on the exterior of a building will not require to be weatherproofed if all the pieces fit snugly together. Not only rain, but wind and cold, have a phenomenal propensity for penetrating the most infinitesimal of chinks.

Putty (also sometimes called 'cement', though this word is also used in stained glass to describe the technique of bonding pieces of glass together with epoxy resin or 'cement') consists of whiting, vegetable black (which helps to give the leading a blackish finish. If you would prefer a bronze patina finish, then substitute red ochre for the vegetable black), powdered red lead, and turpentine (or linseed oil) all mixed together to form a sealant. The process of applying it to a window is by its nature a messy one and, as such, one of the least pleasant that we must undertake; so it is best to lay out some newspapers on your work surface and to put on old clothes before you begin.

First prepare the putty by mixing the ingredients together, adding more whiting if it is too liquid or more turps if it is too solid. The object is to force the puttying solution under the leaves of the lead came tightly up against the glass. For this you will need two scrubbing brushes and a sharply pointed stick, such as an ice-pick. Using one of the brushes, rub the putty over one side of the window, forcing it under all the lead flanges. Then sprinkle an extra handful of whiting over the window. This helps to dry the putty quickly. Now flatten the edges of the leads with a blunt, smooth instrument, which forces the leaves down firmly against the putty. Excess putty will ooze out of the leaves and so now take the sharply pointed stick and run it around all the edges to cut the putty off flush with the leads. Sprinkle another handful of whiting and scrub clean the window with the other brush. This is done because whiting also helps to remove any grease and dirt from the glass and to give it a bright polish. Repeat the process on the other side of the window, and let it dry for several hours. Check, then, whether any extra putty has oozed out of the joints and, if necessary, re-use the pointed stick.

Photograph showing the old technique of using circular steel rods for reinforcement. The window itself is of incidental interest as it is an original Tiffany and includes his unique 'drapery' glass, which is used here to form the white vestments on the figure. Drapery glass consisted of thick, three-dimensional folds of glass resembling those on drapes; hence its name
Reproduced by courtesy of Joni Meyer, Manhattan, New York

Reinforcing

Any window of dimensions greater than about 3 ft square will require structural support every 2 ft or so. This can be done by one of two ways; either by the use of the steel core leads described in chapter 4, or by means of reinforcing rods, called saddle bars, that are fitted to the back of a window after it has been assembled and soldered, prior to its being placed in its predetermined window space.

The choice of which method to use is a personal one, though the former tends to be used more in smaller windows and the latter in larger ones; the reason being that saddle bars provide greater protection against wind and pressure over larger areas than do steel core leads.

Saddle bars have traditionally been made of large, circular steel rods that, because steel is an alloy that is unsolderable, have to be fastened to the leading by means of small lengths of copper wire, called *ties*, that are wrapped around the saddle bar to secure it. There are, however, also galvanized bars on the market, and these, because of their zinc coating, can be soldered directly on to the window, thereby allowing for a neater and perhaps stronger union.

When adding saddle bars to the completed window, make certain that they extend from one border – either horizontally or vertically – right across to the opposite edge, and that, in doing so, that they are soldered not only to each border, but at regular intervals to the leading in the composition. This will greatly increase the structural support that the bar affords the glass. Try, in addition, as mentioned in chapter 7, to incorporate your reinforcing requirements into the design itself wherever possible, by bending the rods to meet the curvature of the leads, thereby disquising their presence. Such pliancy in rods is limited, of course, by the very nature of their structural function, but they can be bent to follow certain contours and angles.

Rubbing the putty solution under the leaves of the lead cames with a brush

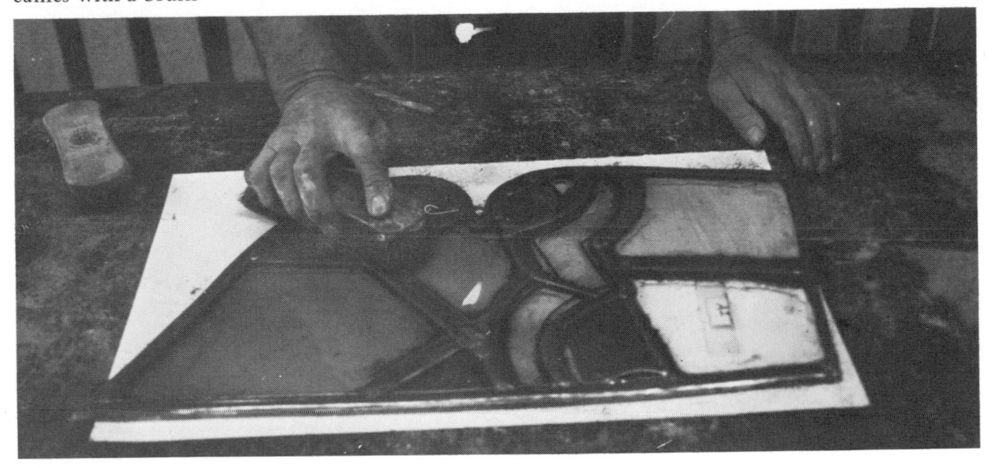

Using a sharp point to remove excess putty. Run the point along the edges of the lead

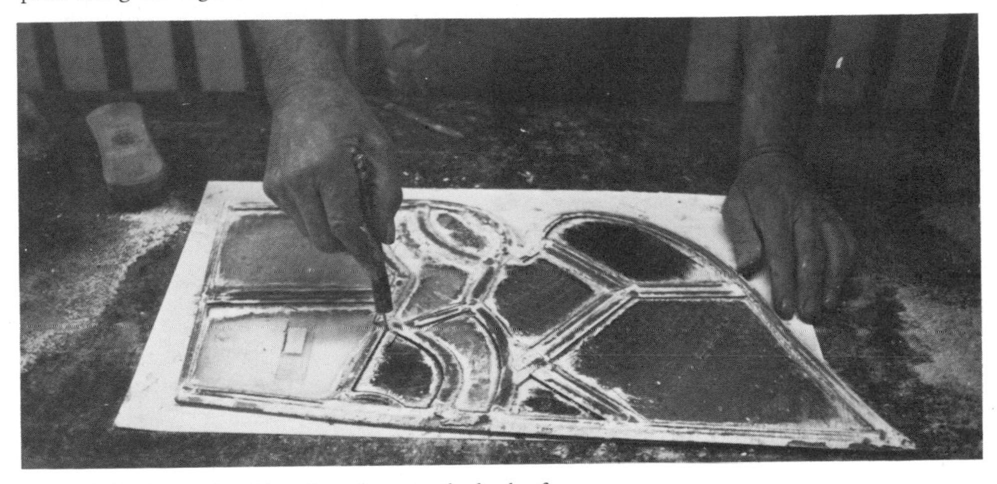

Zinc reinforcing rods soldered neatly on to the back of a window so that they follow the contours of the lead lines in the design

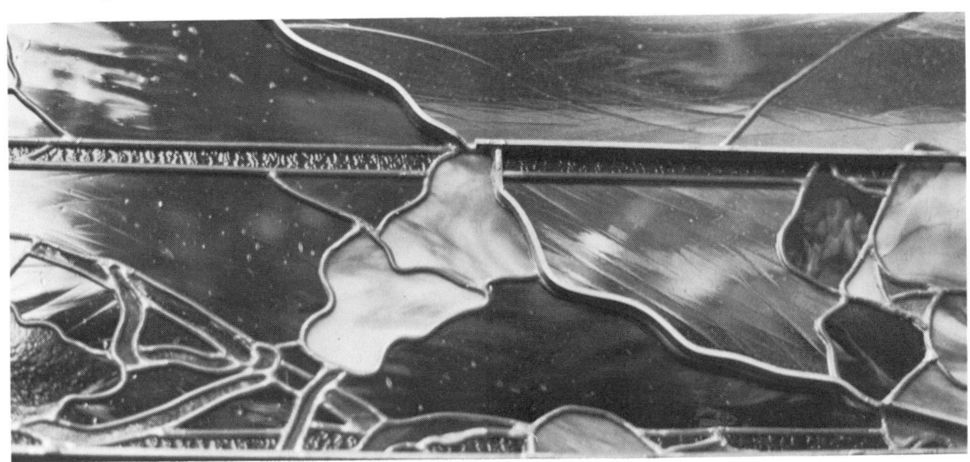

The remainder of this chapter illustrates, by means of photographs, just how wide a range of artistic effects can be achieved in window-form. Coloured glass is used nowadays in windows not only in its traditional role as an ecclesiastical art form, but in any number of other non-religious ways to create windows that are as diverse in style and design content as they are beautiful.

Circular geometric window
Clive Blewchamp, Toronto, Canada

Oak tree with sun and moon eclipse series. 36 in. diameter window containing coloured glass and two-way polarized tinted mirror glass
Dan Fenton, Oakland, California

Another portrait window. No painting; only leaded glass. On the left, three of The Beatles (Paul McCartney is on an adjoining window); on the right, Elvis Presley
Jeff Speeth, Friendship, New York

Triptych window by Anita and Seymour Isenberg, Norwood, New Jersey

Abstract window
Al Lewis, La Fayette, California

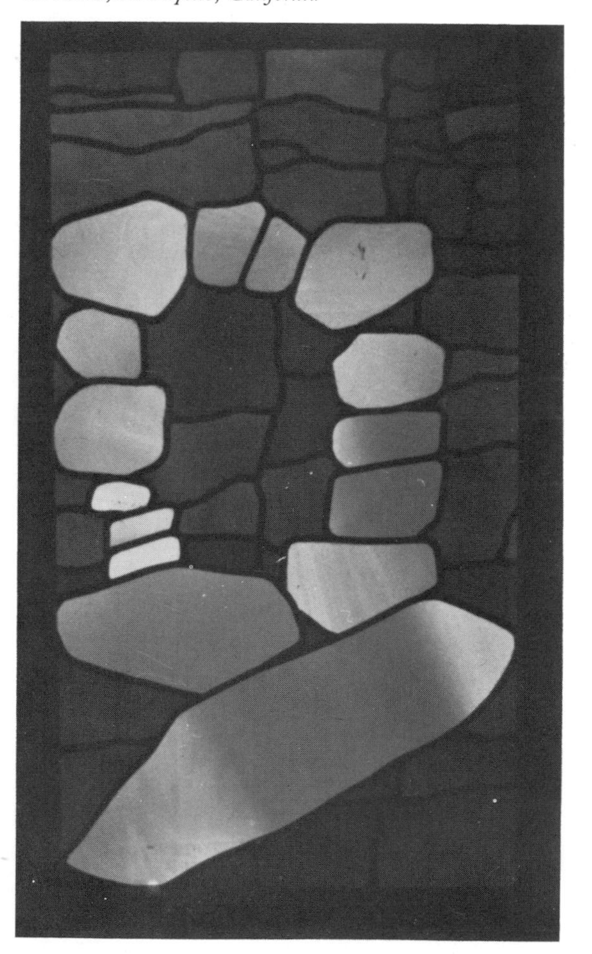

Abstract window
Al Lewis, La Fayette, California

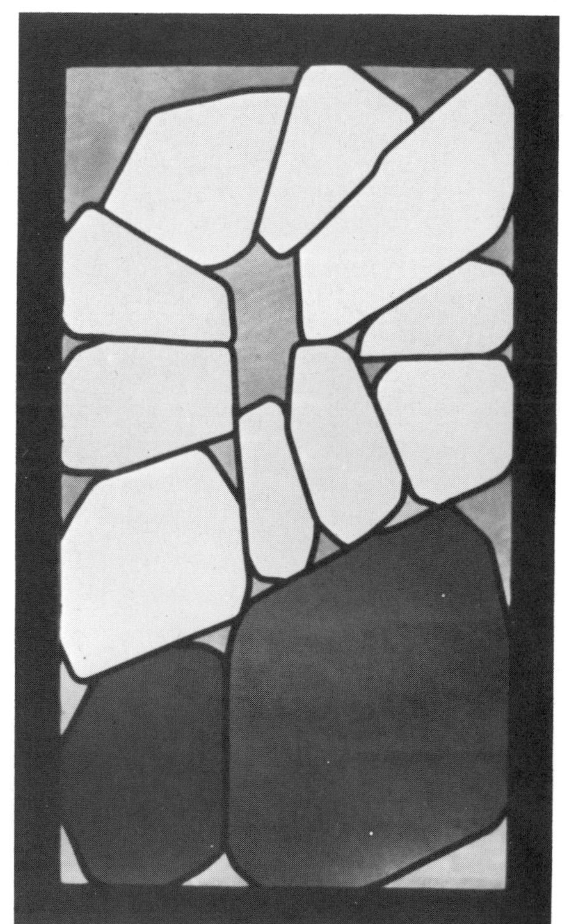

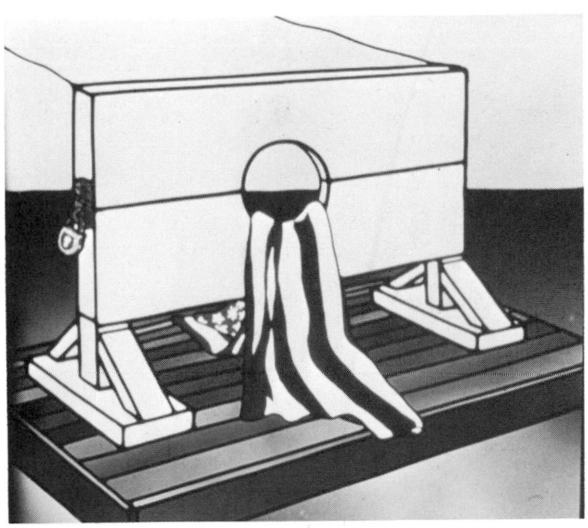
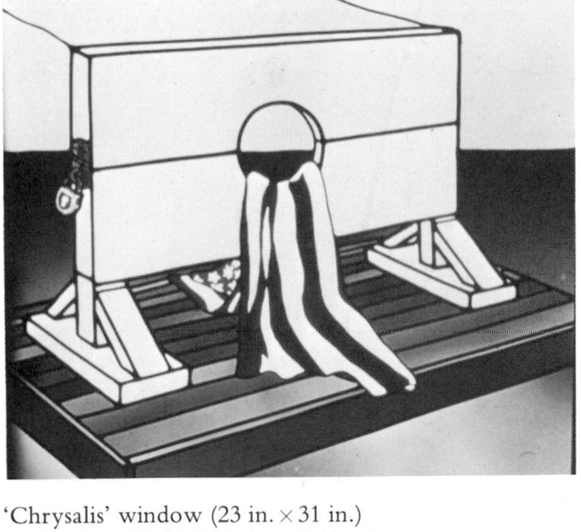

'Monetary Gain' by Paul Marioni, Mill Valley, California. Paul is one of the most versatile of present-day glass artists, using various sandblasting and photo-silkscreen techniques, examples of which cannot, however, be shown in this book as it covers leaded, unpainted glass only

'Chrysalis' window (23 in. × 31 in.)
Kathie Bunnell, Mill Valley, California

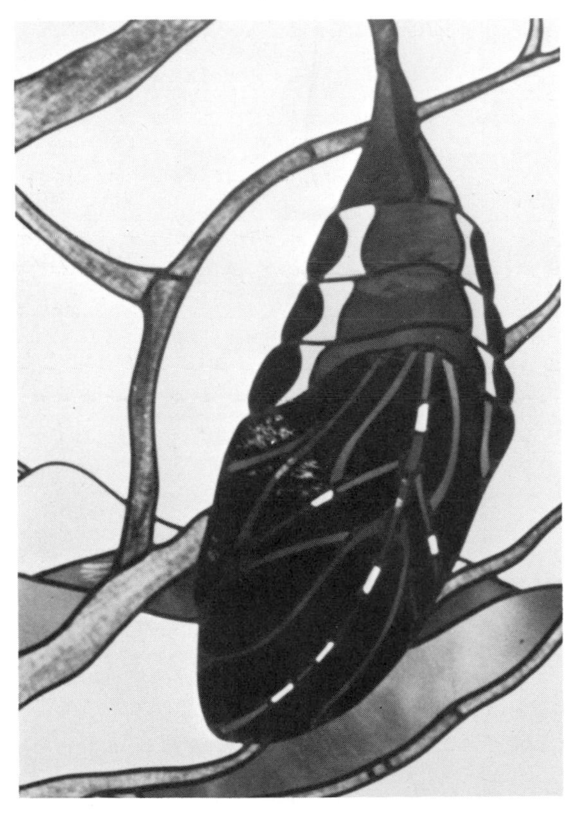

Window by Robert Sowers, Brooklyn, New York, showing the use of glass jewels within a design

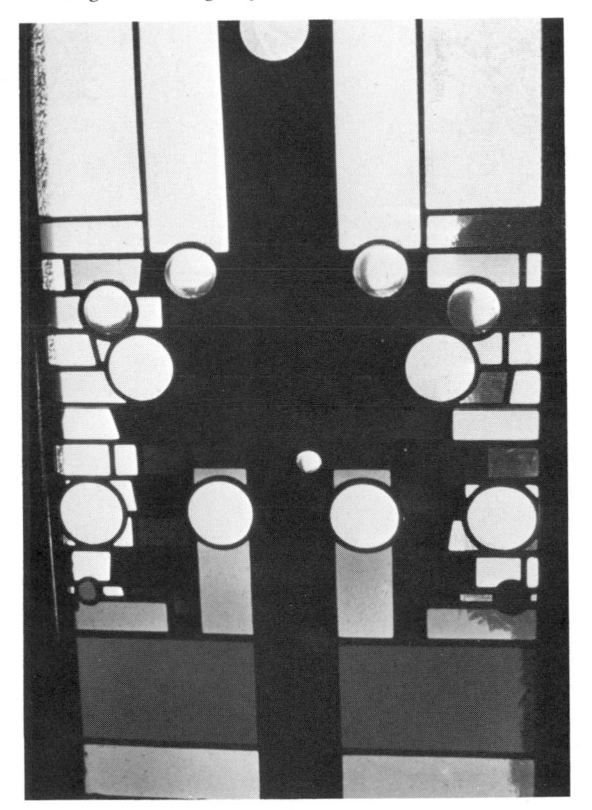

Right
Church window in Our Lady of Perpetual Help, Atlanta, Georgia
David Wilson, Brooklyn, New York

Architectural window
David Wilson, Brooklyn, New York

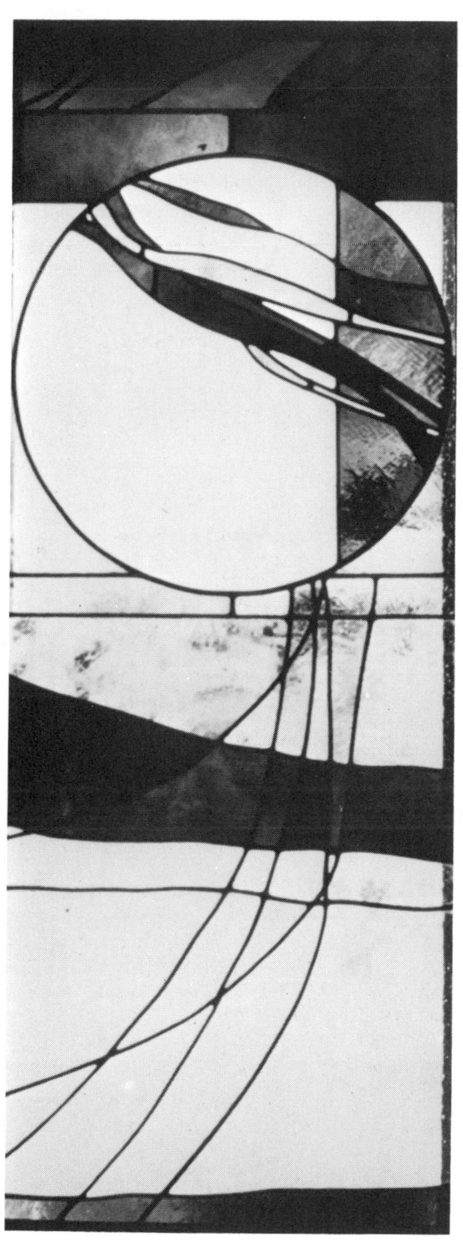

Domestic window
Ed Carpenter, Milwaukie, Oregon

Section of the northside nave of St Marien, Dortmund, by Johannes Schreiter, 1971–2. There are three main colours of glass: three whites; a rust-coloured red; and a sparkle of selenium red

Window with abstract design for Protestant church in Laufenburg, Germany. This shows how contemporary German glass designers have broken with the traditional neo-gothic concept of religious glass. The window contains, in addition, four different sizes of lead width
Johannes Schreiter, 1969

Window in the Temple Sholom, Greenwich, Connecticut
Robert Sowers, Brooklyn New York

Stairwell window by Ray Bradley, London

One of a series of ten symbolic windows in St John's Evangelist Lutheran Church, Holbrook, Long Island, NY, depicting the life of Christ. This one is the Last Judgement
Designed by Dietrich Spahn, Minneapolis, and glazed by Helmut Schardt, East Northport, New York

Door panel by Ed Gilly, Greenwich Village, New York

Peacock window in clear and opalescent glass. The bird's eye is a glass jewel
Reproduced by courtesy of the Light Opera, Ghirardelli Square, San Francisco, California

Window by Peter Mollica, Berkeley, California

Detail from south wall, St Johann, Dortmund-Kurl, Germany
Ludwig Schaffrath

Chapter 9

Lead came lampshades

So far in this book, lead came has only been incorporated in flat projects, such as panels and windows, but it can also be employed in the construction of lampshades and other three-dimensional objects (for the latter, see chapter 13).

One of the most intriguing aspects of leaded glass, and one which for the initiate is often so difficult to comprehend, is that a great deal of three-dimensional glass objects can be assembled in flat form and then bent into their finished shape. This both eliminates the need for a mould and allows extremely rapid and effective construction of different shapes and sizes of lampshades. This chapter describes the method of assembling one such shade, and you will find, as you familiarize yourself with the various techniques involved, that these are fundamental to a multitude of more spectacularly shaped lamps.

The following are guidelines that should be followed, not only in the lamp design in this chapter, but in the construction of any lead came lampshade.

Cutting glass panels

Use a pattern piece, as described in chapter 6, to cut out the panels for the lamp. When the dimensions of each panel of glass have been determined, then transfer these carefully on to pattern paper and cut them out with standard household scissors. Pay particular attention to accuracy in doing this as each pattern piece will be used to cut out 10 or 12 identical pieces of glass, and, obviously, if the dimensions of the pattern piece are initially incorrect – however minimally – this error will compound itself with each additional panel to the extent that, if the error is sufficiently great, the lamp's final shape can be totally distorted. Remember, in this respect, to check the edge of the pattern piece for signs of buckling, fraying and corrosion (if you are using a lubricant with your cutter), as these will also lead to incorrectly shaped panels of glass. Replace the pattern piece when necessary.

If the colours for the lamp have been chosen from irregularly shaped offcuts of glass left over from previous projects, then shift the pattern piece around on these pieces of glass to ensure that as many panels as required can be cut out of them with the minimum of glass wastage. Remember, while doing this, to ensure that, if the glass has a textural pattern, the panels are cut so that the grain in each one runs in the same direction.

If, however, the panels are to be cut from a large sheet of glass, then this should be done as shown in the diagram overleaf to ensure maximum glass economy. First, cut off a long rectangular strip of glass of the same width as the length of the panels in the lampshade, and then, by reversing the pattern piece after each cut, cut out the required number of panels of that colour.

Break off each section as it is cut, ie, before re-positioning the pattern piece, and finally groze away any splinters of glass that may remain along the edges of the panels. Now groze the corners of all the panels before assembling them. Rounding off the sharply-angled corners in this manner allows the panels to be bent more easily without bringing tension to bear on the glass which would, in all likelihood, crack it if such pressure points were not removed. Grozing allows both sides of a project to be soldered before bending it into its finished shape. If, however, you find that your panels do not bend easily after you have soldered both the front and back of the project, then do not solder the back (which will become the inside of the lamp) till you have completely assembled the lamp: at which stage then solder the inside joints. This should not be necessary, however, if you both groze the corners of the glass and bend the panels slowly and evenly.

Choice of lead came

In the lampshade described in this chapter recommended types and widths of lead came are given. These are my preferred dimensions for the design at hand, and not necessarily those that you would choose, and you can, of course, change them. There is, in addition, the possibility that your materials supplier will not have the recommended sizes in stock, in which case it will be necessary to improvize by using the nearest available size of lead. If, for example, you cannot obtain both $\frac{1}{16}$ in. and $\frac{3}{16}$ in. U-frame leads (for the borders on the lamp), then use whichever of the two is available. Remember, when making a choice between various lead sizes, that the selection should be governed by consideration of both the necessary structural support that the lead must provide for the glass and its visual heaviness in the project as a whole.

Simple, yet effective lead came conical lamp, which is made flat then curved into its finished shape

When cutting a series of panels shaped such as 'A' in the diagram, reverse the pattern piece after each cut, to ensure that as many panels as possible are cut out of the sheet of glass. Break off each piece as you go along, and then re-position the pattern

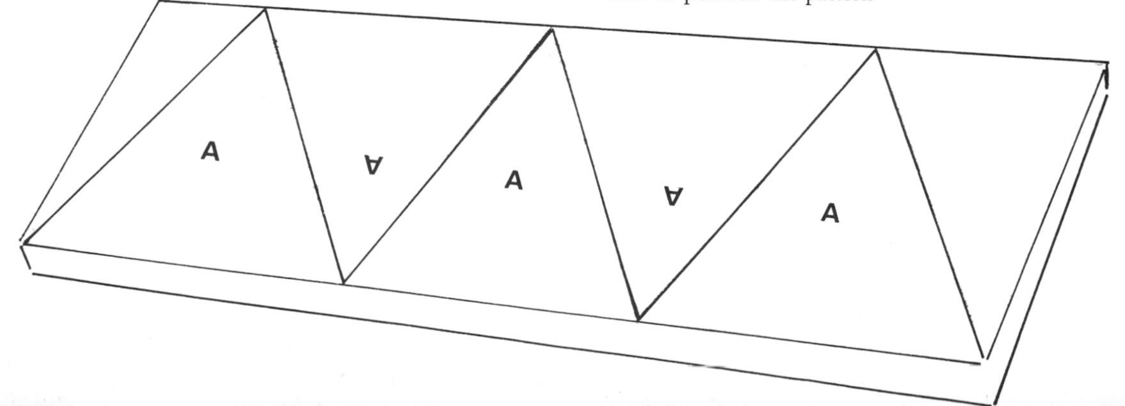

Size and number of panels

The dimensions of the panels in the following lampshade can, of course, be varied to suit personal needs. You should, in fact, always try to determine before you start to make a lampshade, where you will finally put it. This will help you to decide not only which colours to use for the panels, but also what size they, and therefore the completed lamp, should be. So where, for example, you would prefer to make your panels size 10 in. instead of the $7\frac{1}{2}$ in. given in the text (see page 160), then you can do this. Remember, though, that when panels have a crown or skirt, or both (as in the project in this chapter), that you must change the dimensions of *all* of these proportionately. So if, in this case, you change from a $7\frac{1}{2}$ in. panel to a 10 in. one, then the sizes of the crown and skirt must be made 25% greater (4:3) than those given in the text to ensure that they are kept in proportion.

Do not fall into the trap of assembling a lamp out of the first beautiful sheet of coloured glass that you pick up and then try to make it blend, both in size and colour, into the existing decor in your home. It will probably disrupt what was initially a well-balanced interior design composition.

You can, in addition to varying the size of the panels, also use more or less of them than are used in this project. The less panels (of the same size) you use in, for example, a cylindrical lampshade, the sharper will be the angles between the panels and the smaller the diameter of the finished lamp, while the less you use in a conical one, the steeper the angle of the shade. Experimentation will show the various decorative effects that can be obtained in this manner.

Choice of glass

A lampshade, or any three-dimensional glass object that relies on an artificial light source placed inside it for its illumination, should, I feel, be constructed with, preferably, opalescent glass, or the darkest variety of transparent glass obtainable within the colour range that is required in the design.

A naked light source, such as a bulb, is, in general, too harsh on the eye (how much so depends on the bulb's wattage). The more opacity in the glass between this light source and the viewer, the less evident, therefore, the bulb, and the more restful the subsequent effect. Whereas a lamp's primary function is to generate light, such a function must always be conditioned, as far as the glass artist is concerned, by whether or not, in providing this light, it enhances, rather than detracts from, the innate colour and textural beauty of the glass. If the light is too strong it will literally burst through a piece of glass with such blinding velocity that the glass itself is not seen, but only the bulb. Ideally, the light rays should, on hitting the inside surface of a lampshade, be reflected downwards while at the same time highlighting the textural qualities of the glass. Lamps are not supposed to be beacons of light, but rather a subtle blending of functionality, ornamentation, and refinement. They should glow rather than dazzle.

Lead came lampshade with crown, filigree collar, and repeating design on the lower panels
Peter and Joanne Nervo, Berkeley, California

Cupola-shaped hanging shade with a dogwood flower on trellis design.
Alastair Duncan, London

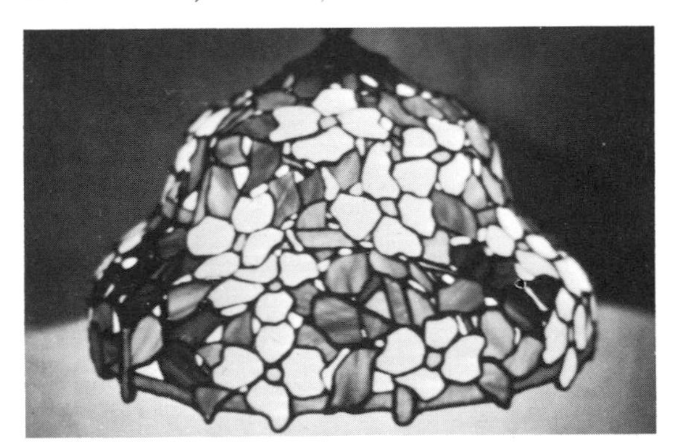

Cementing

Cement will not normally have to be applied to the leads in a lead came lampshade (in the same way as that described for weatherproofing windows in chapter 8) unless the lamps will be positioned outdoors, such as porch or garden lanterns.

Cementing will, however, always supply added strength and stiffness to the skeletal function that the lead came affords the glass, and you can, therefore, apply it if you wish, though this would only be structurally desirable with exceptionally large and heavy lamps. Cement can, in addition, be used remedially to seal off any chinks of light that are visible between badly interlocking pieces of glass and lead came, though in such cases it is always best to replace the offending piece or pieces of glass if you detect the error before you have soldered the lead joints. Remember, in this respect to always press the leaves of the lead down flat against the surface of the glass with a lathekin or other blunt object as this will not only give the lead a tighter grip on the glass, but also minimize the likelihood of the light finding any gaps.

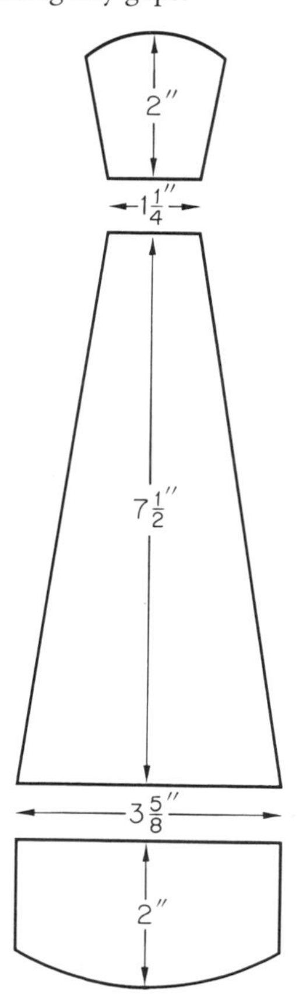

Art Deco lampshade

I have termed this an Art Deco shade as its geometric shape and style are reminiscent of that era, but it can equally well be described as modernistic. It can be used both as a hanging and as a table lamp, though it is better suited to the former.

The lamp consists of three sections: a crown, a body, and a skirt.

Dimensions
Number of sides: 12
Top diameter: $8\frac{1}{4}$ in.
Bottom diameter: 14 in.
Overall height: $10\frac{1}{4}$ in.
Glass Three different pattern pieces are required to cut out the twelve pieces for each of the lamp's three sections.
1 The crown 2 in. deep.
2 The body $7\frac{1}{2}$ in. deep.
3 The skirt 2 in. deep.
Lead You will need three strips of $\frac{1}{4}$ in. round H-frame lead, two strips of $\frac{3}{16}$ in. U-frame, and one strip of $\frac{1}{16}$ in. U-frame. Stretch all the strips, ready for use.

Assembly
(a) *The body* This is assembled flat on the work surface. First, cut 12 pieces of $\frac{1}{4}$ in. H-frame lead of length $7\frac{1}{4}$ in. and then knock two nails into the work surface about 10 in. apart, and place one of the strips against them. Position a glass panel into the exposed channel in the lead, and then, by adding lead and glass alternately, build up the body. (Do not forget to groze the corners of each panel of glass before placing it in position.) The leads should not be quite as long as the glass, and should be placed so that a small amount of glass is visible at either end. Tap each piece of glass so that it fits snugly into the exposed channel of the lead, and then secure it in position with a nail at top and bottom. Make certain, while doing this, if more than one colour of glass is being used, to keep to the predetermined colour scheme. The twelve panels, when assembled, form a fantail shape on the work surface. The next step is to add the top and bottom U-frame leads. First, the $\frac{3}{16}$ in. one for the bottom border. Starting at one end, remove the nails one at a time and fit the U-frame along the entire edge of the body, panel by panel. Make sure, while doing this, that you push it up firmly into the joints between each panel so that it touches on the ends of the $\frac{1}{4}$ in. H-frame leads. This will ensure a neat joint when

Position the first panel into the first $7\frac{1}{4}$ in. strip of lead

The twelve panels of the body of the lamp assembled and held in position by nails

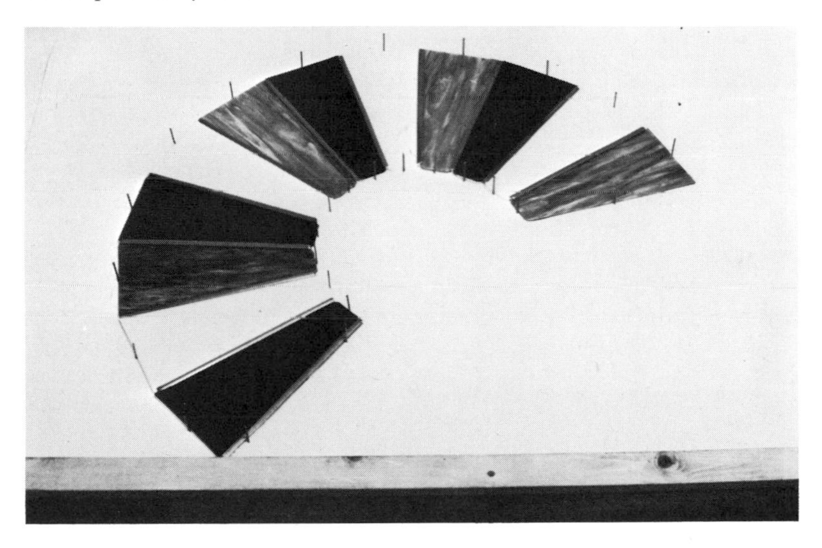

you solder it. When you have bordered each panel, then secure the U-frame lead with a nail. Proceed in this way till you have completed all twelve panels. Then cut the U-frame lead off square with the outer edge of the last panel of glass. Repeat the operation for the top border of the body, again with a $\frac{3}{16}$ in. lead. All except the twelfth panel are now laid out so that they are housed on both sides with H-frame leads and, on their top and bottom edges, with U-frame leads; all pressed firmly together in a complete whole and held in position with nails. It is only the outside edge of the twelfth panel of glass that is exposed: this will fit into the empty channel in the first strip of lead that is put down when the panels are bent round to form a truncated conical shape.

Check, finally, before soldering, that the strips of $\frac{1}{4}$ in. H-frame lead between each panel are the correct length. If, on the one hand, they are too short, then the gap will have to be bridged with solder; if, conversely, they are too long, then they will not house the panels of glass tightly enough and the glass will tend to pop out of the lead channels when they are bent into the final shape of the lamp.

Soldering Now solder all the joints as described in chapter 4: mill the surface of the lead came at the joints to remove the oxide; apply flux; and solder neatly. Make certain that solder does not run between the lead and the glass as this can cause difficulty later in bending the panels. Be careful, also, not to melt the strips of lead when soldering the joints. If you do then replace the damaged lead immediately, ie, before proceeding to the next section. This is as much a decorative necessity as a functional one: not only will a burnt joint be structurally weak, but it will look shoddy, and if you procrastinate – intending to replace it later – then you will probably never get round to doing it as it is infinitely more difficult to do at a later stage, because it will then be necessary to cut or melt the offending lead out of an already soldered area. All reparatory work should be done immediately an error is made.

When you have soldered the top side, remove the nails, and turn it over and repeat the process on the reverse side (you can, as mentioned, leave the soldering on the reverse side till you have assembled the lamp, though this should not be necessary if you have grozed the corners of the panels).

Next, grip the top rim of the panels and raise them up, while bending the two end panels around in an arc towards each other. Keep the bottom edge on the work surface to act as a brace, being careful, when raising the top of the lamp, that the twelfth panel, which has no outside lead strip, does not slip out of place. If it does, then re-position it before continuing. The bending process must be done slowly and evenly to ensure that the panels of glass do not pop out of the lead channels, and also that undue pressure is not brought to bear on the top corners of each panel of glass that might cause it to crack. When you have bent the panels right round into an umbrella shape then fit the exposed edge of glass on the twelfth panel into the empty lead channel on the first one. Then solder the top and bottom joints on both the outside and inside of this last joint.

(b) *The crown* Assembling this entails exactly the same technique as that for the body. Cut twelve pieces of $\frac{1}{4}$ in. H-frame of length $1\frac{1}{2}$ in. and assemble the pieces of glass (reminder: groze the corners of each piece) for the crown flat on the work top, securing each piece with nails at top and bottom as you proceed. Then add a $\frac{1}{16}$ in. U-frame border lead around the outside edge and a $\frac{3}{16}$ in. one around the inside. Solder the front and back and bend the two ends round towards each other, using the work top as a brace. Fit the exposed edge of the twelfth piece of glass into the empty channel of the first piece of lead that you put down, and solder the two joints.

(c) *The skirt* Take two strips of wood and nail them into the work surface so that they abutt at right angles. Then place a long strip of $\frac{3}{16}$ in. U-frame along the inside edge of the wooden strip nearest to you so that its one end fits flush into the corner formed by the two strips of wood, with the flat end, ie, back, of the came pressed against the wood and the open end facing outwards. Next, cut twelve pieces of $\frac{1}{4}$ in. H-frame of length $1\frac{1}{4}$ in., and position one of these along the inside of the other strip of wood so that it abutts the U-frame in the corner. Then, starting from the left, assemble the skirt by placing a strip of lead and then a piece of glass alternately. Make certain, when placing the first piece of glass into the corner formed by the two strips of lead, that it fits snugly into both channels. Tap it, if necessary, with the back of the lead knife or stopping knife. Then secure it in position with a nail. Add the strips of lead and pieces of glass till all twelve of each are in position, securing each in turn with a nail. Then working again from

Fitting the $\frac{3}{16}$ in. bottom border and then the top $\frac{3}{16}$ in. U-frame border *(below)*

the first piece of glass, remove the nails one at a time and add the strip of $\frac{1}{16}$ in. U-frame as a border along the exposed edges of the pieces of glass. Work section by section; completing each one before proceeding to the next. Press the lead round the curved edge of the glass and push it up firmly into the gap between each section with your fingers or with a lathekin so that it touches the H-frame lead. Then secure it with a nail and proceed to the next piece of glass. Finally, cut the end of the U-frame lead off flush with the outside of the twelfth skirt section, and make a critical examination of all the joints to determine whether the leads fit neatly, and replace any torn or badly fitting ones. When you are satisfied, square off the skirt by trimming any remaining pieces of lead came that extend beyond the ends with a lead knife. Then, removing the nails one at a time, solder all the joints.

Bending the skirt

Grip the skirt on each side, and then, using the work surface as a brace and starting at either end or in the middle, bend each skirt panel round *slowly* and *evenly* so as not to crack the glass. Position your thumbs on each pair or joints and bend the panel several degrees before proceeding to the next joint to do the same. Do this gradually and systematically till the panels have been bent right round into a cylindrical shape. Use the utmost patience and caution here because undue haste and carelessness can cause the glass to crack; thereby entailing the long and tedious and often heart-rending process of replacing the panel. This is, for the beginner and veteran glass practitioner alike, an arduous and temper consuming operation; one which is invariably a result of bad workmanship rather than defective materials. When you have bent the panels then fit the outside edge of the twelfth panel of glass into the empty H-frame channel on the first strip of lead that you put down. The glass should fit readily into the channel in the lead, but, if it does not, then check if the channel is crushed and, if so, prize it open with a lathekin. When the glass is firmly in position and the twelfth panel is square with the first one, then mill the surface of the lead at the top and bottom joints, add flux, and solder.

The next step is to solder the three sections together, for which you will probably need an extra pair of hands to help you to hold the sections in position as you first tack them together and then apply solder smoothly over the joints (remember to mill the surface of the lead and to flux it) to give them a neat, finished appearance.

First, solder the skirt to the body – inside and outside – making sure that the panels match, both in size and in your intended colour scheme. Place the body on the skirt and match the two by lining up one of the vertical strips of lead on the body with a corresponding one on the skirt. Then tack the joint temporarily and, moving around, a panel at a time, solder each joint. It is at this stage that you will discover whether or not, when cutting out the sections of glass for the body and the skirt with the pattern pieces, you did so accurately. If you did, then the panels will match perfectly; if not, then the alignment between the vertical leads on the body and skirt will be increasingly off as you move around the lamp.

Next, in the same manner, add the crown. This last operation can be both difficult and messy, as the joint between the body and the crown is recessed and therefore hard to get at with a soldering iron. You may find, as you try to bond the crown to the body, that the joint thus formed is humpy and unsightly. Such blemishes will, obviously, mar the total effect of the lamp. To have completed all but this last step correctly and then to have the overall effect spoiled by the inaccessibility of the last joint to be soldered cannot, simply, be contemplated, and so, if you are dissatisfied with the quality of this joint between the body and the crown, then there are two ways in which you can hide it.

The first method is to take a strip of $\frac{3}{16}$ in. U-frame lead of sufficient length to wrap around the joint, and then to flatten it out on the work top by pushing a blunt object, such as lathekin, into the channel of the lead to prize it open till the lead is completely flat. Smooth it out by pressing down along it until you have ironed out any unevenness. You are now left with a long, flat band of lead. Wrap this tightly around, ie, over, the soldered joint between the body and the crown and then cut it off so that the two ends of the band fit flushly together. Remove the excess piece of band and solder the joint neatly. This will hide any unevenness in your initial soldering.

The second method employs the same principle, except that instead of using a strip of flattened U-frame lead came, a filigreed band of copper is used. The technique of fitting this is described in chapter 12.

Bending the body panels into a conical shape before soldering the remaining unsoldered joint

Assemble the pieces of glass for the crown, ensuring that the intended colour scheme is adhered to

After adding the $\frac{1}{16}$ in. U-frame lead on the outer edge, then fit the $\frac{3}{16}$ in. one round the inside by first removing the nails and then re-positioning them to hold the lead against the glass. Make certain that the H-frame leads are neither too long nor too short for neat soldering

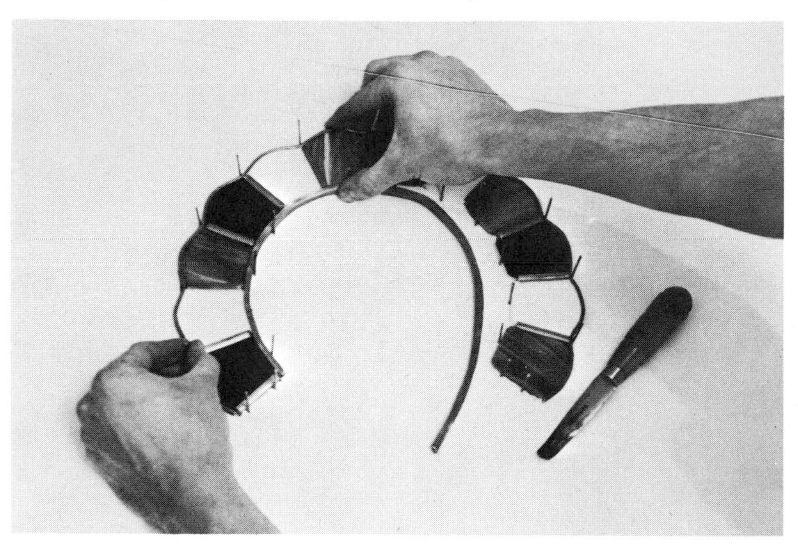

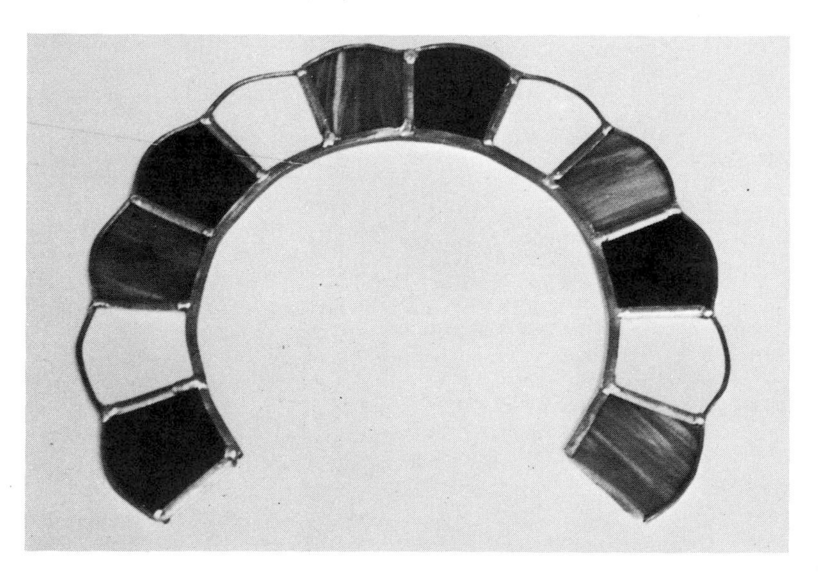

The completed crown, soldered and ready to be bent into shape

Bend the sections around and fit the two ends together before soldering this last joint

Placing the first of the twelve pieces of glass for the skirt into the corner formed by the long strip of $\frac{3}{16}$ in. lead and the $1\frac{1}{4}$ in. H-frame one. Adding each section, while ensuring that the colour scheme is adhered to as planned

Assembling the panels for the skirt

Soldering the top $\frac{3}{16}$ in. U-frame lead to the $1\frac{1}{4}$ in. H-frame leads

Attaching the lower $\frac{1}{16}$ in. U-frame border of the skirt

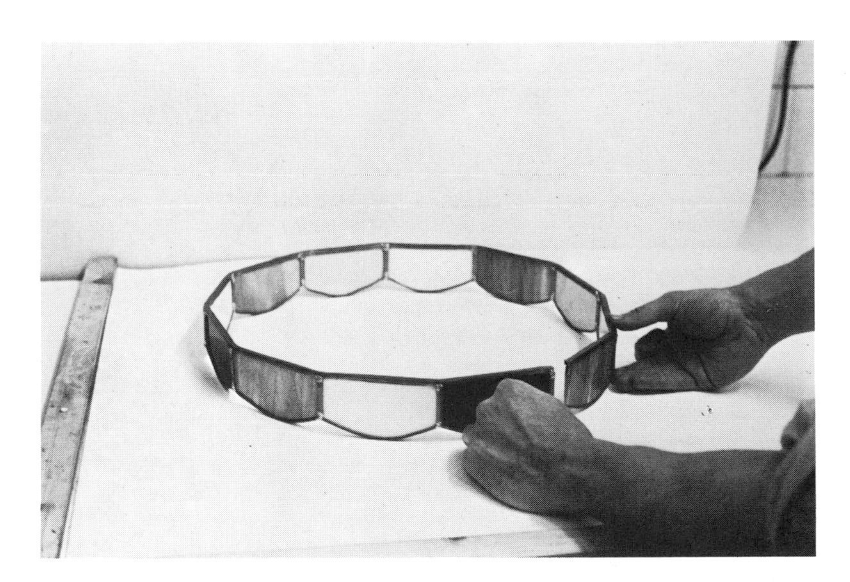

Drawing the pieces around into their polygonal shape

The three completed sections, ready for final soldering together to form the lampshade

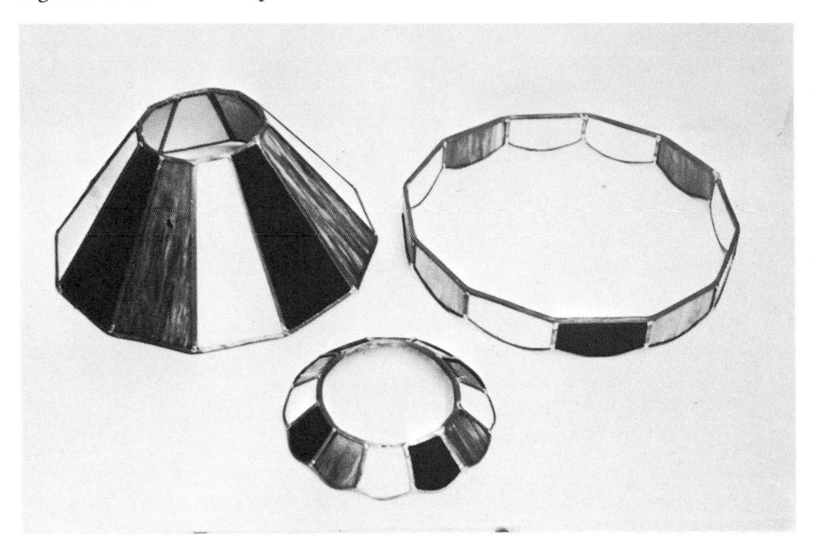

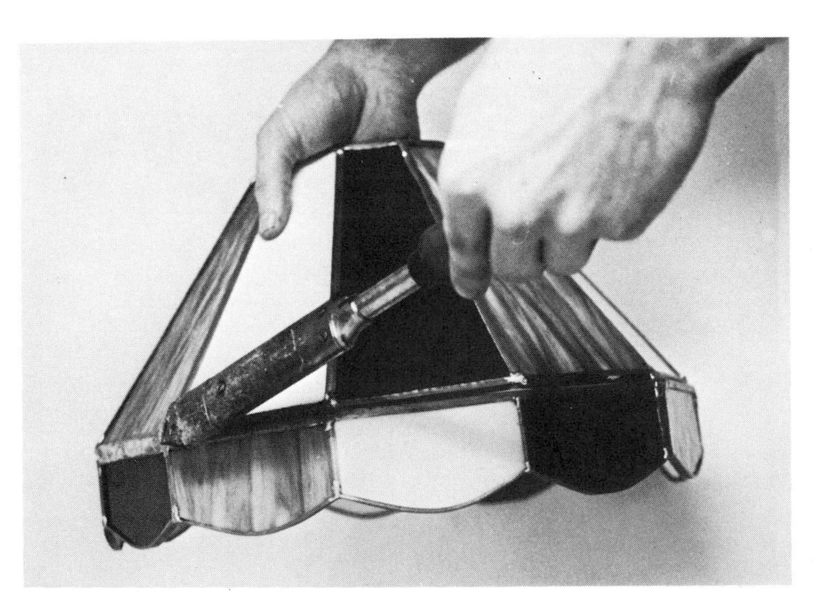

Once the body and skirt are tacked, then solder each joint between the body and skirt neatly

When you have soldered the crown to the body in the same manner as you soldered the skirt, then touch up all joints to ensure that they are tidy

Finishing

All that remains is to scrub the lamp thoroughly with a liquid cleaner to remove all the flux and incidental grime, and to darken the soldered joints with a blackening agent or patina, if you so choose; to add a spider to the top to support the lamp; and to wire it. The last two operations are described in chapter 12.

The completed lampshade, with collar and spider attached ready for hanging

Unusually shaped lamp in lead came
Reproduced by courtesy of Mugworts, Berkeley, California

Differently shaped lead came lamps
Nervo Studios, Berkeley, California

Lead came, tiered lampshade in which the bottom panels have a repeating floral design. It is assembled flat on the work surface and later bent into its final shape. Note also the crown and filigree collar
Joanne Nervo, Berkeley, California

More lead came lamps
Nervo Studios, Berkeley, California

Summary

Lead came or copper foil for lampshades

With experience, it will be possible to determine whether lead came or copper foil is best suited to the shape and style of lampshade you are setting out to construct. The following are two potentially limiting aspects of lead came that should be born in mind when making your choice:

1 The sharpness of the angles in the design. For every width of H-frame lead there is a certain maximum angle that it will allow beyond which the edges of the glass will pop out of its channels. For wider leads this angle is, obviously, greater than for very narrow ones where the leaves of the lead are too short to grip the glass effectively other than in a flat plane. So, when deciding on the design for a lampshade in which the glass is housed in lead came, it is important to consider both the width and angle of the lead joints that you want in the design. It is a good general rule, until you have worked with several widths of lead and have learnt what angles they will tolerate, to choose the wider one when deciding between two lead sizes.

2 As a general rule I would suggest that any three-dimensional object which can be assembled flat in panel form on the work surface and then bent into its final polygonal shape should be made with lead came, while any object that requires a mould, such as a cupola- or dome-shaped lampshade, should be made with copper foil. Alternatively, you may decide that not only panelled lampshades, but also geometrically patterned curvilinear ones made on a mould, are better done with lead came than with copper foil. The choice is a personal one and will vary depending on the specific design requirements of the lamp you are assembling. Do not, though, consider using lead for floral-styled lampshades; it really cannot compare aesthetically with copper foil in this instance.

Three-tiered lead came table-lamp, with a scalloped skirt
Joanne Nervo, Berkeley, California

Chapter 10

Copper foil

All the projects up to this point, both two- and three-dimensional, have used the traditional lead came technique of bonding pieces of glass. Leads perform this function satisfactorily except in two instances: when the design requires thin, sharply-angled joints or very fine detail. In the first case, the narrower the lead came the less acute the angle it will allow in the joint, as the edges of the glass will pop out of the channels in the lead; while, secondly, the inherent bulkiness of lead came prevents it being used effectively where the design calls for small, clearly delineated areas of glass to provide precise and intricate detail. In such cases lead came swallows up the glass and gives a weighty, non-decorative appearance.

Copper foil is an alternative method of interlocking pieces of glass that not only overcomes these two problems, but can substitute for lead came in any given instance. The process consists of wrapping the pieces of glass in copper foil which, when soldered, holds them together. All the techniques that you have learnt so far in this book, eg, the method of soldering or cutting glass, apply exactly as they do for lead came: all that is happening is that one method of housing the glass is being replaced by another one. Copper foil is, in fact, possibly the easiest and most alluring of all leaded glass techniques; one that though well established in America, has only recently been introduced into England.

As a material copper foil has paradoxical qualities which are marvellously suited to our needs: whereas it is light and flimsy and malleable, it provides, when soldered, absolute structural rigidity; being able, for example, to support the weight of hanging glass objects far better than lead. The reason for this is that copper foil is soldered along its entire length, and not only at the joints, as with lead came. It does, in addition, allow the sophisticated glass practitioner tremendous versatility in his choice of materials: not only flat glass, but anything that can be wrapped in copper foil can be incorporated into the design. Pebbles, sea-shells, glass jewels, cabochons, porcelain – to mention just a few of the type of ornament that Tiffany and other Art Nouveau exponents used with such magical effect – can be interposed with the glass in the design.

Copper foil was first incorporated into glass work at the turn of the century in America, most notably, of course, by Tiffany, and it is supplied commercially today in two forms: either as an adhesive-backed tape or in sheet form.

Adhesive-backed copper foil

Copper foil tape can be bought in rolls 36 yd in length and of various widths, the most suitable for our needs being $\frac{3}{16}$ in., $\frac{1}{4}$ in., and $\frac{3}{8}$ in. The tape has an adhesive backing which is, in turn, itself backed by a protective strip of paper or plastic which must be peeled off as the tape is used. The tape provides, due to its uniform width and ease of application, a very fast and neat means of wrapping pieces of glass in copper foil.

Which width of tape to use depends primarily on the thickness of the glass used and, to a lesser extent, on the artistic effect that you wish to achieve. First ensure that the tape is wide enough to overlap the edge of the glass and to grip its top and bottom surfaces tightly. Once the minimum width of tape required for the piece of glass that you are using has been established, then a larger size will both increase the solidity of the joint and make it look more bulky. For example, a $\frac{3}{16}$ in. wide tape allows thread-like, wispish detail and intricacy of design, while a $\frac{1}{4}$ in. tape would, on the same glass, be more substantial, both in strength and visual effect.

Application of the adhesive-backed copper foil
Check, when all the pieces of glass have been cut and are ready to be assembled, that the copper foil is wide enough to fit around the edge of the thickest piece of glass to be used. In the same way that, when using lead cames, it is important to ensure that the heart height is large enough to house the thickest piece of glass, so too with copper foil. Different widths of tape can, of course, be interchanged just as can lead cames of different heart heights. But make certain before starting a project that the tape, or tapes, available are sufficiently wide to wrap *every* piece of glass. This is particularly important when using handmade English antique glass as its wavy surface varies considerably in thickness along its entire length.

I recommend starting with $\frac{1}{4}$ in. copper foil as this is a good general purpose size, being wide enough for most varieties of glass. (If, incidentally, when using $\frac{1}{4}$ in. foil you would prefer a narrower width, it can be reduced by cutting away the unwanted section with a razor blade or scissors.)

Check, finally, that there is no oil (from the glass cutter) or dirt on the edge of the glass as such impurities can prevent the glue on the foil from adhering properly. If necessary, wash and dry the pieces.

The next step is to wrap the foil.

1 Peel the protective paper off the copper foil, and then, taking a piece of glass, place its edge along the middle of the strip and wrap the strip around the entire edge of the glass. Be careful to keep it in the middle of the foil so that there is an equal margin of overhang on either side. Then overlap the start of the foil slightly and either cut off the excess tape with a pair of scissors or break it carefully with your fingers.

2 Crimp the foil around the edge of the glass, making certain that it does not crinkle on the corners.

3 Lay the glass down flat and press the foil tightly on to its top surface with a blunt instrument such as a lathekin or flat piece of wood. Then turn the piece of glass over and repeat the operation on the bottom surface. Check, while doing this, that the overlap on each side of the glass is even in width. If not, then cut away any irregularities with a razor blade or a utility knife. (This is usually more successful than trying to pull the foil off the edge of the glass to re-position it once it has been pressed down, as the foil grips with a tenacity which is amazing for a material that is so seemingly lacking in body and strength.)

4 When satisfied that the foil forms a neat, uniform border to the glass, then start on the next piece and continue till they are all wrapped. The next step is to solder.

Two rolls of copper foil. The one on the left is of $\frac{3}{16}$ in. width, the other $\frac{1}{4}$ in.

A recent acquisition to the glass artist's range of potential materials: slices of rock, in this instance Brazilian agate. Rocks can be cut into slices – a width of about $\frac{1}{8}$ in. is ideal for copper foiling – with a lapidary saw and then polished, if need be. These slices can then be further cut into more manageable sizes and shapes, such as circles and ellipses, for incorporation into lamps and windows. Agates such as onyx and quartz are particularly suited to our needs due to their crystalline properties and translucency
Obtainable from Nervo's, Berkeley, California

An example of what can be done with copper foil . . . and imagination. Pebbles of various hues wrapped and soldered together on a mould to form a lampshade. The effect is one of soft, yet rich, luminosity
Joni Meyer, Manhattan, New York

Another example of copper foil combined with creativity. Lamp consisting of, amongst other things, a car head light, optical lenses, glass jewels, roundels, circles of glass, all wrapped in copper foil and soldered together
Ed Gilly, Greenwich Village, New York

Copper foil in sheet form

The use of copper foil in sheet form for leaded glass originated at the turn of the century in America. It has only been very recently that copper foil has become available in tape form with an adhesive backing. Despite the rapidity and ease of operation that the latter affords, the sheet form variety of copper foil can produce work of exactly the same calibre, but with the disadvantage that some preparatory work has to be done on it. It does, though, have two advantages which should be taken into account before you begin a project.

First, cost. Sheet form copper foil is, because it is not precut or adhesive-backed, far cheaper than copper foil in tape form. You will have to decide for yourself whether the time spent on cutting the sheet into strips in which to wrap the glass is offset by the money saved in not buying the tape. I find that the decision depends, finally, on the volume of work at hand. If the project to be done is a small one, then it is a simple matter to cut the strips of foil manually. If, however, you are embarking on something more grandiose, such as a lampshade containing perhaps 1000 pieces of foiled glass (a not, in any way, inconceivable number of pieces as you will find as you familiarize yourself with the medium and as your ambitions grow accordingly), then, in this case, you will probably prefer to forgo the money saved, but time spent, on wrapping a vast amount of strips by hand.

The second advantage that sheet form copper foil offers over tape is that it can be cut to accommodate any thickness of glass required to be incorporated in the design. This applies, in particular, to handmade sheets of glass such as English antiques that can vary considerably in thickness along their edge. In such cases a uniform, say, $\frac{1}{4}$ in. tape can be both too wide or too narrow (ie, overlap the glass by too much or not at all) at various stages along the same piece of glass. In order to overcome this you can measure the thickest part of the glass, cut a strip of foil off the sheet wide enough to cover this, and then, when you have wrapped the glass, cut away any unevenness in the overlapping edge of foil with a utility knife or razor blade.

You can also, in this way, include within the design any glass or non-glass three-dimensional object that can effectively be gripped by a piece of foil. Not only roundels, bullions, glass jewels, and pieces of dalle de verre, but any light-reflective or refractive ornament that can be wrapped in foil and that will enhance the total decorative effect, is an artistically legitimate component of your work. Measure the width of the object and cut a piece of foil to house it.

Copper foil can be purchased in sheet form in various sizes and thicknesses that do not have adhesive backing. The length and width of the sheet does not matter as it will be cut up into thin strips, but its thickness does. The reason for this is that the thicker the foil the less flexible and nimble it is, and the less easily, therefore, that it will mould itself against the contours of the edge of the glass. I use a 0·016 in. foil which is suitable for most foiling purposes. Thinner sizes offer increased flexibility but less strength and rigidity, while thicker ones offer the opposite.

Various methods of preparation and application of sheet form copper foil exist, of which I would recommend the following:

First, take the sheet and cut it into workable sizes, such as 6 in. × 12 in. Then tape one of these by its ends to a piece of thick paper. Mark the sheet off lengthwise with a ruler in sections of the width of foil, eg, $\frac{1}{4}$ in., that you require, and, taking a razor blade or utility knife, cut out the $\frac{1}{4}$ in. strips, leaving them taped to the paper. In this way they will not get enmeshed with the work in progress, and can be filed away neatly in an office file holder until required.

Remember to check the thickness of the various sorts of glass in the project to determine the different widths of foil that will be needed, and in each case to allow sufficient width of foil to cover the edge of the glass and to overlap the top and bottom surfaces.

When you have cut out what you estimate to be a sufficient number of strips for the project, you are then ready to use them to wrap the pieces of glass. Fold the foil around the glass in exactly the same way as that described for the adhesive-backed tape, and when you have overlapped the start of the strip, then tear off the excess tape and smooth out the joint. Now flux the joint and apply a minute blob of solder to it to bond the two ends together. Finally, crimp the foil firmly against the glass all around the piece. The foil should now hold the glass securely until the project is assembled. Follow the same procedure for all the pieces of glass and then position them on the work drawing or mould prior to soldering.

If you experience difficulty in getting the strips of foil to grip the glass satisfactorily, and that even when you have soldered the ends of the foil, that the glass still pops out, then you can apply your own adhesive-backing to the foil to help it to stick better. This is best done when the strips of foil are still taped to the paper (see above), as trying to wrestle with foil and adhesive on the work bench can indeed be a sticky business. Either rub the strips with vaseline or spray them on one side with 3M permanent adhesive, and, when the application becomes tacky, then wrap them around the glass and solder the joint as before.

Wrapping the foil around the glass. Make certain that the glass is in the middle of the foil

Using sheet copper. The sheet is cut into lengths of the desired width, which are then wrapped around the edge of the glass. The point where the two ends meet is then fluxed and soldered to hold the foil securely on the glass

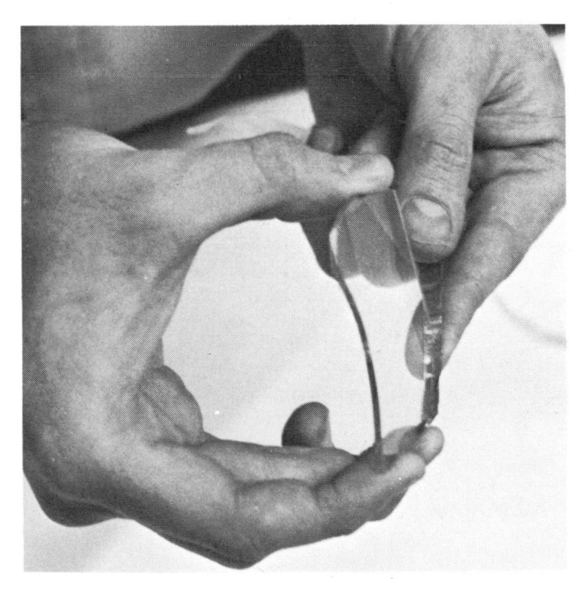

Crimp the foil in to both sides of the glass as you proceed

A roll of sheet copper

Tinning

This entails applying a thin layer of solder on to the outside edge of copper foil. First the foil must be fluxed, and then solder is melted on to it and smoothed out to form a thin, even layer. 'Tinning' can be used for our purposes to give a rigid, uniform finish to any exposed edge of foiled glass, eg, the lower border of a lampshade. Instead of using a U-frame lead for the border you can simply tin right round the entire foiled edge of the shade (the process strengthens the foil, enabling it to hold the glass firmly in position). Tinned foil does not readily give as straight a border as U-frame lead, but there may be instances – for example, some floral lampshades look very attractive with a wavy or jagged bottom rim – when a tinned edge is perfectly suited to supply the required effect.

Soldering

When all the pieces of glass have been wrapped in foil they are ready to be put together and soldered.

If the object you are making is flat, such as a panel or window, then this should be done on a work drawing in the same way as described in chapter 6. If, though, the object is curvilinear, such as a lampshade, then this entails the use of a mould. When assembling two-dimensional objects place all the foiled pieces of glass in position on the work drawing to ensure, before they are soldered, that they fit correctly and do not extend beyond the outside perimeter. Check, if the pieces of glass extend beyond the lines on the work drawing underneath them, whether such discrepancies are due to incorrectly cut glass or because the foil has not been pressed tightly against the edge of the glass. Where, however, gaps exist between the assembled pieces of glass, you must decide whether one or all of the pieces involved should be re-cut and re-foiled or whether such gaps can be plugged with solder without reducing the required sharpness of detail in the design.

When making a three-dimensional object then this must be done on a mould. Soldering pieces of glass on to curvilinear surfaces entails the same process as that used on flat ones, although when pieces have to be soldered on an angle then they must be secured with pins or by 'tacking' (see below) before the complete soldering operation is undertaken. When you are satisfied that the pieces fit correctly then you are ready to solder them together.

First, apply flux to all the pieces and then 'tack' them. 'Tacking' entails bonding the pieces of glass provisionally by melting a blob of solder on to each foiled joint. This holds the whole project together until and while you are soldering it properly, which is a necessary precaution as the pieces tend to slip when being soldered. Make certain, though, before tacking, that the glass is correctly positioned, ie, flush against its neighbour, because to try to 'un-tack' tacked pieces of foil can be an exasperating business: the result invariably being that the two will only separate by tearing the copper foil off the edge of the glass.

When all the pieces have been tacked together, then, starting in one corner, systematically solder the entire project. The solder should adhere readily to the copper – bonding the foiled edges of the adjacent pieces of glass instantaneously. Apply the solder until all the copper foil has been covered.

Preparing to assemble a triangular panel. Each piece of glass is wrapped in copper foil and put on the work surface prior to being positioned in its place on the work drawing

Assembling the pieces on top of the work drawing

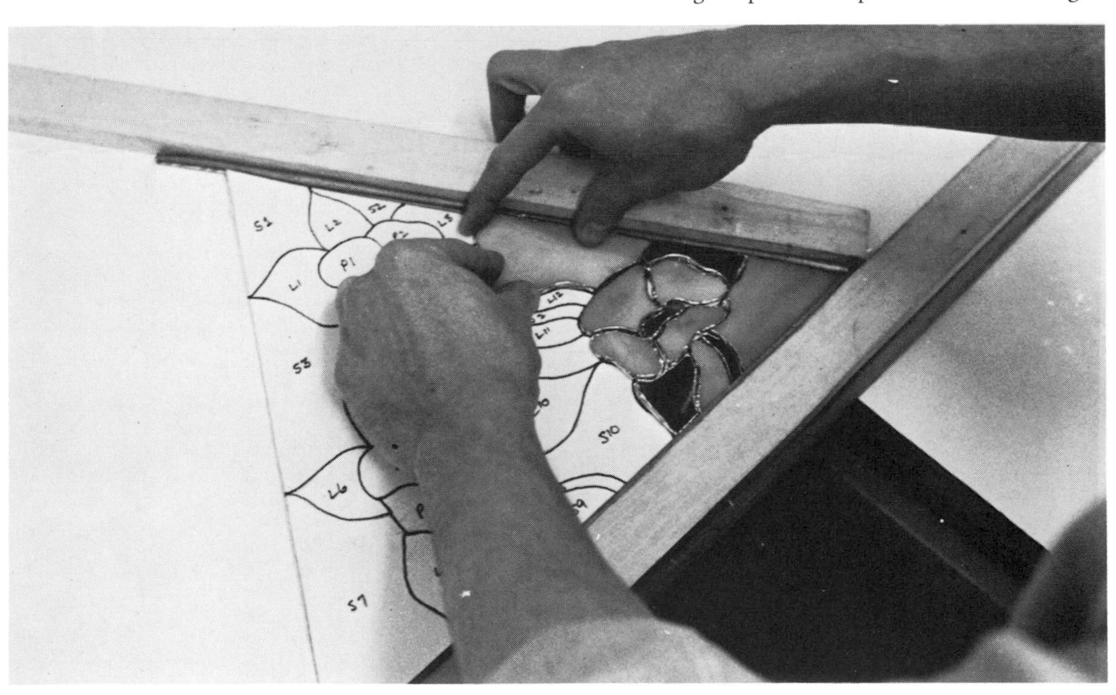

Hold the iron so that its bit is horizontal, ie, flat, then use it to shape and steer the solder as you melt it on to the joint.

The size of the soldering iron bit depends on personal preference, but as the average width of joint created by two pieces of foiled glass in juxtaposition is usually very narrow, so most glass practitioners prefer a small bit, such as $\frac{1}{4}$ in. for copper foiling. Large sizes of bit tend both to obscure one's view of the soldering operation and be too bulky in hard-to-get-at corners inside three-dimensional objects.

If you find that the solder tends to drain right through the edges of the copper foil and solidify on the underside of the joint, and that, when you turn the glass over to re-solder it, that it promptly disappears back through the edges to the other side, then the soldering iron is probably too hot. Unplug it and let it cool, or, if a variable resistor (rheostat) is attached to the iron, lower the temperature by means of the dial. Conversely, if the solder will not flow smoothly along the foil, but drags, forming small peaks and valleys, then the iron is too cold. You will be amazed at the mercurial characteristics of the solder as you run it along the foil. It forms waves that flow back and forth, only solidifying when you remove the soldering iron. You can, in this way, keep re-fashioning it till you are satisfied with its shape. Add more solder until it builds up into a uniform, convex border. Then turn the object over and solder the back exactly as you did the front.

The process of building up the solder is generally called *beading*, and is a good measure of the degree of professionalism that a leaded glass exponent has acquired in his work. Well-beaded copper foil is uniformly rounded throughout the project. There should not, as occurs in soldering lead cames where the solder is only applied at the joints, be any visible distinction between the edges and the joint. The solder should form a continuous whole. Whereas this entails the use of more solder than is structurally necessary (thereby entailing additional financial outlay for solder), it is a decorative prerequisite. The best beading results are achieved by reducing the heat of the iron so that the solder is in a paste form, neither completely solid or liquid. In this way it can be shaped and re-shaped with the bit until all unevenness disappears. If the bit is too hot the solder will disintegrate into myriads of particles that fly off over the surrounding glass.

When the soldering is completed, make one final, meticulous examination of all the joints to ensure that the solder forms a uniform, convex ridge on both the front *and* back, or, in the case of lampshades, on both the outside *and* inside of the object. All soldering, regardless of whether it is readily visible in the finished object or not, should be thoroughly done. In the same way that an antiquarian will determine as much, in terms of authenticity and quality of craftsmanship, from those parts of an antique that are not normally seen, so an experienced leaded glass artist will always check the soldering on the inside of a lampshade to determine the overall level of virtuosity of its creator.

When satisfied with the standard of the soldering, then clean off any flux and globules of solder that remain on the joints and adjoining glass with a soft brush and cleaning liquid. Next, dry it with a cloth or paper towel and scrape away any particles of dirt in the corners of the glass.

The next step is to give it an antique finish.

Right
When all the pieces have been positioned, then the outside border is attached and secured with nails, and the joints are fluxed and soldered

Problems that can occur with copper foil

1 There can be no margin of error when using copper foil for incorrectly cut pieces of glass; less so, even, than with lead came where the leaves of the came can, on occasion, hide the fact that the glass is fractionally too small. Pieces of glass, when wrapped in foil, however, must lie flush against each other to ensure a strong, rigid joint. Gaps between the foiled edges create problems in soldering as the solder tends to flow right through the hole. Remember, in this respect, that the copper foil technique does not require a $\frac{1}{16}$ in. gap between each piece of glass as must be allowed for the width of the heart of the lead when using lead came. So standard household scissors, and not pattern shears or dual line cutters, must be used when cutting out the pattern pieces when you are going to use the copper foil method.

2 If the piece of glass, when cut, breaks at an angle other than a right angle to its top surface, this will in effect increase the area of glass to be wrapped, and therefore the width of tape needed. In addition, in such a case the copper foil will not get as firm a grip on the glass as when the edge is perpendicular, and it may slip off the top or bottom surface of the glass when you come to assemble and solder it, thereby leaving a hole between the glass and the foil. This, in turn, shows bad workmanship, weakens the joint, and will subsequently need to be plugged with solder to stop the light showing through it.

3 Filling holes with solder, either between the glass and the copper foil or in joints between the foiled edges of adjacent pieces of glass, must be done so that any initial error cannot be detected. If the hole is small then solder on its own will probably plug it effectively, but if the solder drains through the gap then one method of overcoming this is to break off a piece of foil, and fold and wedge it into the cavity. Flux it as normally and then apply solder. This should form an effective bridge which will camouflage any initial error on your part – either by miscalculating the size of the pieces of glass or in applying the copper foil incorrectly.

4 If, while soldering, you notice that the solder will not adhere to a section of the foil and, no matter how much you coax it, that it simply rolls off on to the adjoining glass, then this is probably because the foil is dirty or has formed a protective surface layer of oxide that will have to be removed before soldering. Clean the foil by rubbing it lightly with a wire brush or steel wool, then reflux, and resolder. No section of copper foil, however miniscule, should be visible in the end-product, not only, most obviously, because foil that is not covered with solder is not supporting the glass structurally, but also because copper stands out starkly against a background of solder, giving an unfinished appearance. This is especially evident when you have antiqued the solder as this makes the distinction between the two even more pronounced.

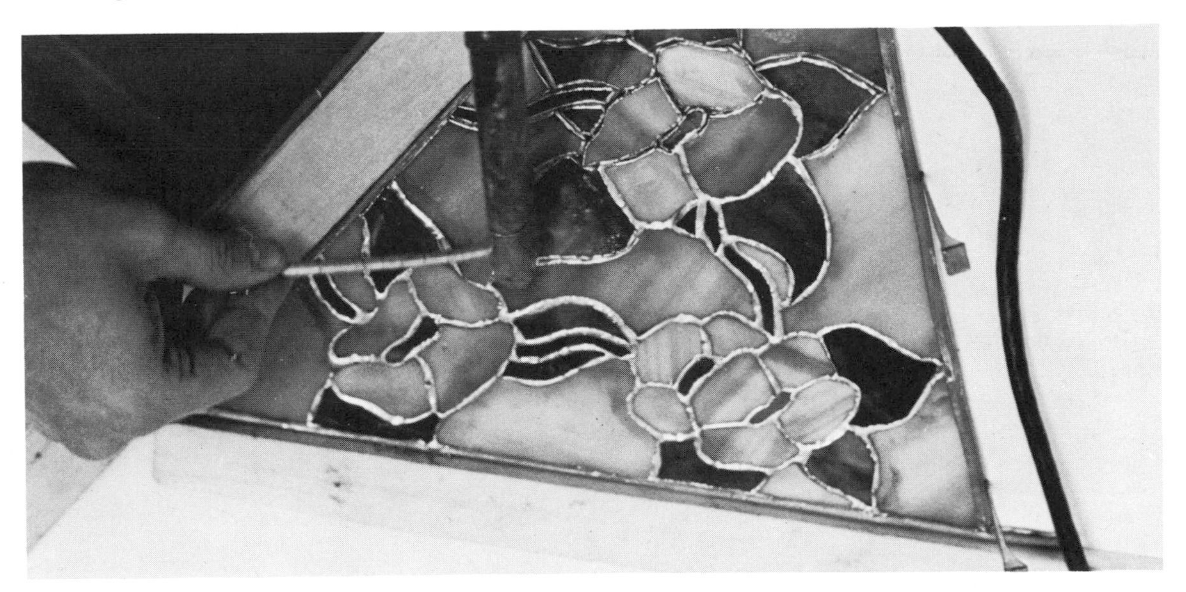

Mixing copper foil and lead came

Mixing lead cames and copper foil within the same project is perfectly feasible from a technical standpoint as the two can form a very functional alliance. The normal procedure, when combining them, is to use lead came for the perimeter and copper foil for the inside leading. The copper foil allows great intricacy of detail within the design, especially on curved surfaces, while a neat, uniform U-frame border gives the object a professional finish. The same applies when making two-dimensional panels or windows: the inside section can consist of foiled pieces of glass and the outside of U-frames or H-frames (remember to use only H-frames when the border will not be visible when finally positioned in the window space).

The two can also, of course, be interchanged within a project as and when you choose. A good rule in this case is to use foil for small, sharply-angled parts of the design where lead came is too unwieldy to follow the contours of acutely shaped pieces of glass, while, on the other hand, lead came is well-suited for those areas that consist of long, straight lines, as the leads provide good linear detail and take less time to assemble in such instances than that spent in wrapping long edges of glass in copper foil.

Whereas there are no technical constraints in mixing lead came and copper foil, there are, however, very definite aesthetic shortcomings of such a union which will now be examined.

Finishing processes for copper foil and lead came mixed

Solder can, by being treated with a blackening agent or a copper sulphate solution (see page 52), be given an antique finish.

When using copper foil to interlock the pieces of glass, it is advisable, when you have finished soldering, to apply either a blackener (for a dark effect), or $CuSO_4$ (for a bronze patina effect), to the solder to highlight its skeletal function in relation to the glass, and to make it appear aged and therefore 'authentic'.

Lead came, on the other hand, will not, unless specially treated, react to the copper sulphate or blackener because it has a protective layer of oxide which effectively prevents the finishing agents from changing its surface colour. So, if copper foil and lead came are mixed within one project, then the end-product will show a noticeable disparity in colour between the antiqued solder (on the copper foil) and the undarkened lead came. No matter how structurally sound and neat the union between the two, the leading will look disharmonious.

The most common instance of this is in the construction of lampshades, where the body of the shade is made of foiled pieces of glass and the edges at the top and bottom are U-frame leads. To eliminate the distinction between the soldered foil and the lead you should mill the lead came along its entire length with a wire brush to remove all the oxide. Then flux it immediately (the flux seals off the de-oxidised surface from the air and acts as a catalyst in the subsequent soldering process), and solder *all* of it, ie, not just the joints, by running the solder along its surface with the soldering iron so that the solder forms a thin, uniform layer on top of every part of the lead came. All the leading in the end-product, both that consisting of lead came and that of copper foil, should now be covered with solder. Choose whichever antique finish you feel is best suited to the style and colours of your project, and apply it evenly with a cloth. The leading should now be uniform in colour and form.

Conclusion

Copper foil gives a leaded glass craftsman vast design potential, enabling him to produce any number of shapes ranging from the flat to the curved to the multi-dimensional. Its powers of malleability and adhesion allow it to fit into any nook or cranny along the edge of the glass and, once positioned there, to cling most tenaciously, providing a firm foundation for the solder.

The choice of whether to use copper foil or lead came is one of personal preference: the former can substitute for the latter in any given instance and it has the advantage, in three-dimensional glass forms, of greater versatility and precision. On the negative side it can take considerably more time to wrap each piece of glass in foil than it takes to cut and position strips of lead came, and it is more expensive to use because the cost of solder is greater than that of lead came (both are sold by weight). You will find as you familiarize yourself with copper foil, that its potential, not only for conventional glass objects, but also for free-form experimentation and glass sculpture, far outweighs any negative features it may have.

The next chapter shows how to incorporate it in the construction of differently shaped lampshades.

Chapter 11

Copper foil lampshades

If you have been to Henry Africa's in San Francisco or to Maxwell Plum's in New York, or to a host of other similarly decorated shops, restaurants, or bars in America in between these two, you will have seen their dazzling displays of glass lamps, some florally and some geometrically designed. Most of these are original Art Nouveau creations, stemming from the early 1900s, with the glass being interlocked by means of copper foil.

The following is the step-by-step procedure of making just such a lamp, as it is a major ambition of nearly every newcomer to leaded glass to make his or her own shade, whether of old or new design – the technique being, of course, the same whatever the chosen style. Chapter 10 described the technique of working with copper foil, and this knowledge is now applied to constructing a lampshade on a mould. An early realization, to the uninitiated, is something that comes as an anomaly: that, whereas the mould is *round*, the glass is *flat*. How, then, are these two seemingly irreconcilable differences to be overcome? (I should mention here that panels of glass can in fact, be bent – the technique is known as 'slumping' or 'sagging' – to follow the contours of a curved surface, but the process is not discussed in this book because it requires the use of a kiln.)

The difficulty is solved by dividing the design linearly so that each individual piece of flat glass will, when assembled, be small enough to conform to the overall shape of the mould. If a piece is too large it will extend beyond the surface plane of the mould at that particular point, thereby breaking the desired smooth outline of the edge of the lamp. There is, in addition, a further consideration concerning the tolerable size limits of each piece of glass: most lampshades are shaped so that the angle of the glass varies in sharpness at different places along its edge; for example, some lamps have a relatively flat top surface that extends outwards for several inches from the central vase cap before turning sharply downwards to form a vertical border or skirt. In such instances the top, flatter surface will allow the use of larger pieces of glass than will the point where the lamp surface turns downwards and is, therefore, more acutely angled. Here, the pieces will have to be much narrower to prevent their jutting out beyond the edge of the mould. Remember, therefore, when making a design, to always ensure that it makes allowance for the angle of the curve at that *specific* point on the shade, with the sharper the angle, the smaller the piece of glass that can be used.

Now, back to the lamp in this chapter , which will be assembled on a Styrofoam mould (you can, if you are making your own mould, use any number of suitable materials such as wood, *papier mâché*, or plastic. It is advisable, though, that beginners buy a completed one from a stained glass supplier).

The major problem in making curvilinear lampshades is that of transferring one's design from sketch form on to the mould itself, with differently shaped moulds presenting different problems. The main difficulty is that the surface (in our case, a hemispherical one) upon which the shade is to be assembled is curved on two planes, ie, both from top to bottom and around the shade. You cannot, therefore, simply draw your design on a flat piece of paper and then attach this to the mould – as you would attach a work drawing to the work surface when making a glass panel – because, whereas it is possible to bend the paper in one direction, it cannot be made to follow the contours of both planes simultaneously.

There are various solutions to this problem, such as, for example, the method of dividing and then cutting the pattern design into segments which are then attached to the mould, but the mechanics involved are invariably complex and need to be specially adapted to the shape of the mould being used, and the simplest and most sensible method for beginners is to draw the design directly on to the surface of the mould itself. The procedure is as follows:

1 Draw the design on to the mould, preferably with a pencil as this is more easily erasable than ink. In any event, you can always paint over or spray a mould to give yourself a 'clean sheet' for the next lamp.

2 I recommend, especially if you are a beginner, that you choose one design and repeat it two, three or more times around the shade. Do not be dissuaded from such repetition on the grounds that it is, *per se*, less than innovative. The most that one can usually see of a lamp's entire circumference at any one time is approximately a half of it, and so a duplicated pattern is not readily identified as such. The advantage of 'repeats' for newcomers is that one does not have to draw on to the entire mould, but only on to a section of it. If, for example, you have a design which you would like to repeat, say, four times, around a circular shade, then you need only mark off 90° on the mould ($\frac{1}{4}$ of the total circumference, which is 360°) with a protractor or set square and then draw your design on to this area, and then do all subsequent working from this. This greatly reduces the amount of time spent on pattern piece numbering, colour-coding and cutting (below).

3 It is a good idea to number and colour-code each piece of the design on the mould, especially if it is extremely complex. This will prevent confusion and loss of pattern pieces when you later cut and assemble the glass.

4 Now transfer the design back off the mould so that you can make the pattern pieces that you will require to cut out the pieces of glass. This is done by placing a piece of tracing paper against the mould and then carefully tracing the design on to it (do not forget the numbering and colour codes). The tracing paper is, in turn, placed on top of a sheet of carbon paper, and the two are then positioned on top of the pattern piece paper, on to which the design is transferred.

5 Using standard household scissors, cut each pattern piece out of the sheet of pattern piece paper. You should, for complete accuracy, make the pieces *fractionally* smaller than the traced lines to allow for the copper foil which will be wrapped around the glass (the space taken up by two strips of copper foil, back to back, is less than the $\frac{1}{16}$ in. heart width of a H-frame lead came).

6 Select the colour scheme for the lamp and use the pattern pieces to cut out the individual pieces of glass accordingly.

7 Using the techniques described in chapter 10, wrap a piece of glass in copper foil and place it in its position on the mould, securing it with two or more pins so that it cannot slip or fall (an alternative method is to use double-sided adhesive).

8 Take a second piece of glass and repeat the procedure. There are no set rules on whether to assemble the glass from the top or from the bottom, or, even, from the centre, moving upwards and downwards. Do whichever you prefer, but I personally always start at the top and work *round and down* centrifugally so that the weight of the glass and the solder is evenly distributed, which is a factor that helps to prevent slippage when you later have to continually position and reposition the mould in order to keep the area that you are soldering horizontal.

9 While assembling the pieces of glass with pins – ensuring continually that they fit snugly and that none of them is overly large so that it juts out beyond the intended overall contour of the shade – you should 'tack' solder the foiled joints between the pieces to keep the composition together and to prevent each individual piece from falling off the mould. Tacking entails dabbing flux on to the joints and then spot-soldering a blob of solder on to some point on each of these joints. Be careful, both while tacking and later when you completely solder all the joints, to ensure that all the pins are removed.

Art Deco type hanging lampshade

An unusual lamp by Duffner and Kimberley that was shown in the firm's 1907 catalogue as 'Dome number 1003. Price $315.00'. Just one of the large collection of turn-of-the-century shades at Henry Africa's bar and restaurant in San Francisco

An old 'fruit' lamp of trellised grapes. Unsigned, probably 1920s

Reproduced by courtesy of Mr & Mrs Brooks, Brooklyn, New York

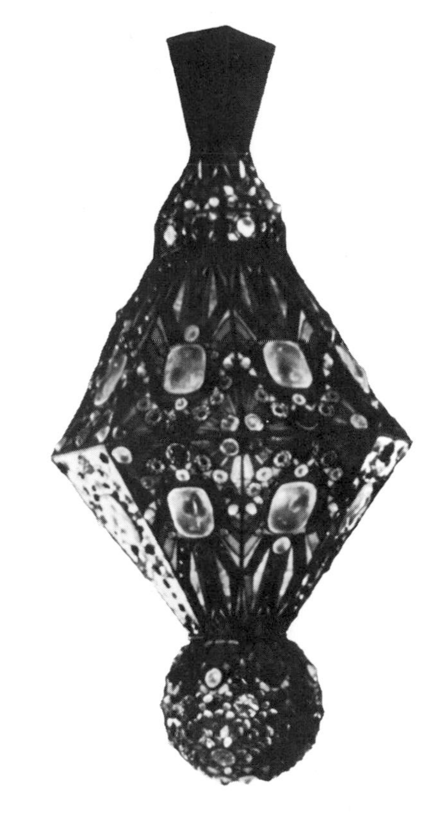

A most unusual lampshade, both in terms of shape and size, comprised of pieces of flat glass, convex and faceted glass jewels, and numerous other glass baubles. The effect is dazzling

Maxwell Plum restaurant, Manhattan, New York

Copperfoiled lampshade by Caroline Swash, London

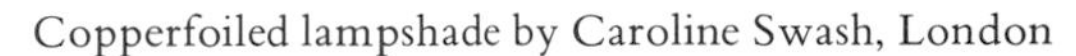

The mould with the design drawn on to it

The first piece of glass in position on the mould and secured with two pins; one at top and bottom

Add each new piece of glass, position it with pins, and then tack adjacent pieces together as you proceed

Proceed round and down the mould. Notice the pins and the tacking and that the pieces are carefully aligned with the design on the mould

The outside completely soldered

Turn the shade over and neatly solder the inside joints as you did the outside

10 Wrap, position, and tack the pieces of glass till they are all on the mould. Now clean, flux and solder all the joints, as described in chapter 10, on the entire outside of the lamp by positioning and repositioning the mould so that the immediate area being soldered is horizontal. When you are satisfied that the outside is neatly soldered, then lift the lamp off the mould, turn it over gently on the work top, and solder the inside in as meticulous a manner as that shown on the outside.

11 Add a top and bottom border lead, such as a $\frac{1}{16}$ in. U-frame lead, to give the shade a neat finish; check thoroughly for blemishes in your soldering; scrub the shade with a brush and liquid cleaner; dry it; and apply an antiquing agent to the leads. Finally, attach the appropriate lamp fittings as described in the next chapter.

The completed shade soldered and patinated and ready to have its light fittings attached (see chapter 12)

Right
Old, unsigned geometrically designed lampshade in copper foil
Reproduced by courtesy of Paul McCartney, London

Tiffany-style geometrically designed shade. Notice the very precise copper foiled leading
Reproduced by courtesy of Light Opera, Ghirardelli Square, San Francisco

Hexagonal flower and trellis lampshade. Each panel is assembled flat and then the six are soldered together to form the finished shade
Tom Rodriques, San Rafael, California

Table lamp by Duffner and Kimberley. The latter initially worked for Tiffany Studios but then defected to form, with Duffner, a company that was operative for about 7 years in the early 1900s. Their lamps are characterized by superb colour harmony and very intricate copper foiling
Reproduced by courtesy of Barry Friedman, Manhattan, New York

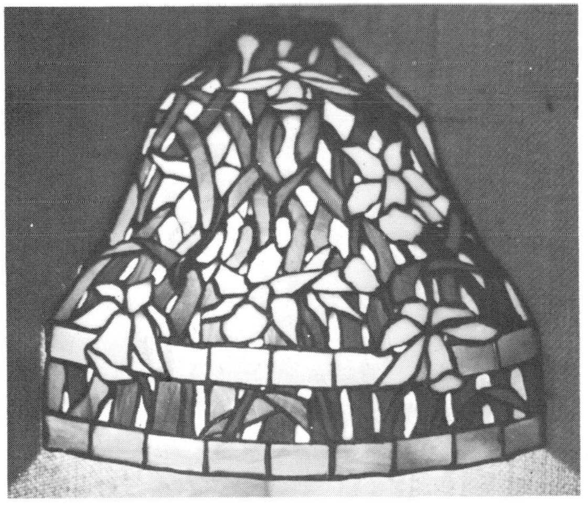

A jonquil corner lamp. Lampshades do not have to be 360°: half a lamp can be made, ie, one of 180°, and attached to a flat wall, or, as in this instance, a 90° one to fit into a corner of a room
Alastair Duncan, London

Accurately and neatly foiled lampshade

Dome-shaped floral lampshade. The flower centres are glass jewels
'Sirius', Swiss Village, Paris, France

Chapter 12

Finishing operation

When you have assembled a lampshade either with lead came or copper foil, soldered it neatly, applied an antiquing agent, and scrubbed and polished both the leading and the glass so that the lamp is sparkling clean and neat, you have arrived, finally, at the last stage; ie, that of finishing it. This entails fitting it on to a base or hanging it; wiring it; selecting your preferred light wattage; and, lastly, signing it.

These last stages are typically the ones to which the least attention is paid by the glass exponent, who, sensing victory, rushes to complete his masterpiece. It is a strange fact that a person who will spend many hours diligently designing, cutting, leading, and soldering a lamp will often sacrifice all the patience and expertise that he has shown up until this stage – in the sudden urgency of trying to finish it – by hanging the lamp just any old how, by fitting the first bulb he finds no matter how powerful or weak it is, and, in general, ignoring all the rules of aestheticism that he applied so conscientiously to assembling the glass. Some people have the feeling that, because they do not themselves 'make' the lamp fittings and bases as they did the lamp itself, that the latter are by this fact somehow outside their control or responsibility, and not an integral part of the final lamp. This is, needless to say, a false premise, as if the lamp is not tastefully displayed, then all the artist's virtuosity has been in vain. The ultimate success of the lamp does, therefore, hinge on the finishing operation. In the same way that an ill-chosen picture-frame can ruin a beautiful painting, so too with lamp fittings – they significantly add to or detract from the end-effect.

Lampshades are generally displayed by one of two methods: either on a base as a table lamp or by a chain or wall bracket as a hanging lamp. The decorative considerations and electrical components required for the two are, in part, the same, but there are certain differences, and we will, for this reason, discuss them separately.

Table lamps

These consist of lampshades that are supported on bases. It is as well to warn you initially that bases are a perennial problem in that they are always in short supply; whereas large numbers were cast in brass and bronze in the early part of the century, they are not so any more. It is a good idea, therefore, to keep a constant vigil for them when you are browsing in antique shops, second-hand furniture stores, and attics, even. In addition, their scarcity is reflected in their price, and they are now seemingly worth their weight in gold rather than brass. I mention this fact to prepare you for the shock that the base that you would like to buy for your shade may well cost more than all the other materials that you have incorporated into it; something which often seems outrageous to a leaded glass artist who is conditioned to thinking of lamp fittings as ancillary to his creation.

There are several considerations to selecting a base. First, and most important, is the size ratio of the base to the shade; the height and thickness of the base vis-a-vis the diameter of the shade. There are no hard and fast rules here other than to try the two together to see whether they 'fit'. Some bases contain a rod and screw mechanism to make them adjustable heightwise, and this greatly increases the range of shade sizes that these bases can accommodate.

Secondly, base style. Here the choice of base is a personal one, but there are certain aesthetic guidelines. Do not, for example, fit an ornate, rococo base to a modern shade as this is obviously incongruous. Try always to match the two by artistic style, yet again, a very fussy base should not be matched with a similarly 'busy' shade as the former will compete with the latter. In such a case the base should be unpretentiously functional to highlight the intricacy of the shade.

Thirdly, base materials. Brass, bronze, and iron are the most common materials for bases as they have the solidity and structural strength needed to support the combined weight of the glass and solder. Bases made of aluminium and other lighter metals or alloys, will topple at the slightest touch.

Fourthly, colour harmony. You should, if possible, match the colour of the base with the leading on the shade. If you have applied a bronze patina to the leading then the base should, ideally, be of a similar colour. Likewise, a black finish on the leading should be complemented by a dark base. An easier way of matching the two is to finish the shade in the colour of the base, thereby eliminating the time and energy to recolour the latter, but if you do need to change the base then there are antiquing solutions on the market, such as patinas and lacquers, which can be applied.

Components of a table lamp

Apart from the base itself, a table lamp consists of the following:

Electrical fittings These usually fit on, or near the top of, the base shaft, and consist of either a single or a cluster of sockets; the manner of attachment depending on how the shade is connected to the base (see below). The number of bulbs to use requires both an artistic and a functional consideration, with a cluster tending to disperse the light more evenly than a single one. It is much better to use two or three bulbs of lower wattage than a

One of the most charming of lamp bases: a graceful 'caryatid' standing on top of a scalloped ash-tray. Aestheticism and functionalism combined. Classically Art Nouveau

Reproduced by courtesy of Henry Africa's, San Francisco, California

single high one as the latter tends to concentrate the light far too strongly on only one area of the shade.

The electrical link from the socket(s) to the source of power, usually a wall-plug, is by means of a cord that passes down through the inside of the base (all lamp bases should contain a central, hollow passage for this purpose) and then emerges near or on the pedestal. Finally, the on-off switch can be placed at various points: on the wall, on the cord, or by means of a chain attachment on certain sockets. Remember, again, to try to choose a cord of a colour which will blend with the overall appearance of the shade and base.

Shade attachment Various bases employ different methods of supporting the shade; the two most common being a harp mechanism or by a rod that extends up from the bottom of the base through the electrical fittings to secure the shade at the centre at the top. The various components – such as screws, nuts, and nipples – necessary to complete the link-up are always carried by a supplier of electrical lamp fittings.

Vase caps, spiders, and finials If the top aperture of the shade is circular then it is usually best secured by two vase caps; one on the inside and the other on the outside of the glass, with the former supporting the shade from underneath and the latter both holding it in place from the top and giving the shade a decorative finish. These vase caps should

An electrical fitting consisting of a cluster of two light sockets. The rod is adjustable heightwise and fits in or on to the base itself

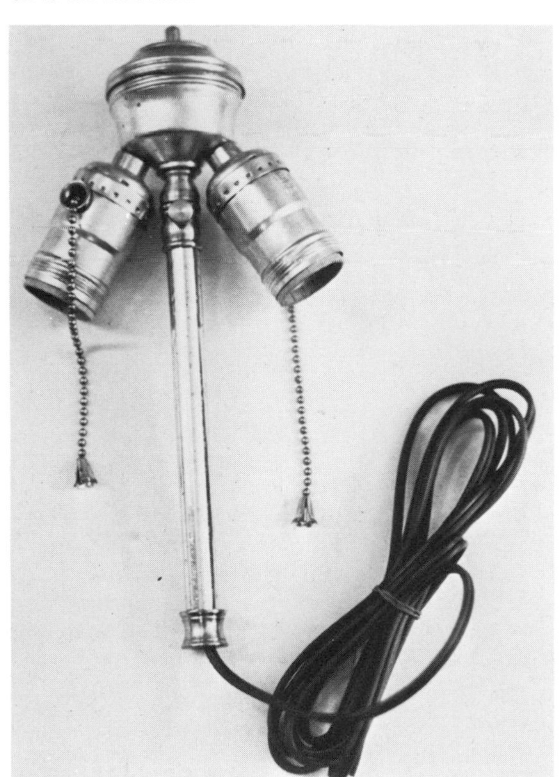

An original, signed, Tiffany base showing one method of supporting a shade. Three arms, on which the shade is positioned, protrude from the central shaft, thereby eliminating the need for an attachment at the top. Tiffany employed the lost-wax technique of casting bronze bases; a process that results in the destruction of the mould every time a base is cast, entailing correspondingly high production costs. Both for this reason and because of the unique patina that he applied to the bronze, his bases are now considered almost as valuable as the shades themselves

A four-armed spider. The arms must be measured and cut off so that they can be soldered on to the top aperture of the shade

Electrical fittings for a hanging lamp: chain, metal loops (only one is needed), two vase caps, nipple, light socket

be of the same size and be only fractionally larger than the opening so that they hold the shade securely while covering as little as possible of the glass that you so laboriously cut and leaded up – the purpose of which is, after all, that it be seen. If, however, the aperture is not circular but, for example, square or polygonal, or if the lamp has a crown, then circular vase caps cannot effectively be used, and you should use some other form of attachment such as a spider or iron crossbar. A spider consists of a central nut from which several arms protrude horizontally. The rod in the lamp base passes through the nut and the arms of the spider must first be measured and then cut off with a saw so that they can be soldered on to the edges of the top opening of the shade.

Lastly, on the very top of a table lamp, comes the finial. This is the nut into which the nipple (threaded pipe) protruding through the vase caps, either on the end of the base rod or on the harp attachment, is screwed to hold the composition of base and shade together. Finials are offered in a range of shapes from a purely functional nut to much more elaborate designs that act as a decorative fillip to the lamp as a whole.

Three different finials, varying in ornateness

A harp attachment. The screw at the top fits through the two vase caps and into the finial, thereby securing the shade

Two bases: one ornate, the other (a copy of Tiffany's Wisteria base) more plain
Obtainable from Nervo Studios, Berkeley, California

Hanging lamps

Components
These vary in some respects from those used in a table lamp, and I will describe them starting from the top and moving downwards.
Ceiling caps These are manufactured in various sizes and designs and are screwed into the electrical box fitting in the ceiling to both neatly cover the aperture where the wiring protrudes from the box and to support the combined weight of the chain and the shade. Ceiling caps are made of brass or bronze and have a hollow hook through which the electrical cord passes and on to which the top link of the chain is attached. In the event that the aperture on the ceiling is not situated directly above where you wish to hang your lamp, then an additional, smaller ceiling cap – in effect, only a hook – should be screwed into the ceiling at the required point, and then the chain is 'swagged' from the larger cap to the hook, from where it hangs vertically. Swagging is not only functional in that the wire, and hence electricity, can be relayed from the central point to any part of the room, but it can be highly decorative, with gentle loops of chain providing the lampshade with very elegant support.
The chain There are any number of chains available, varying in size and design of link. Try, when making your selection, to match the colour and style of the chain with that of the lamp. They should balance.

The one end of the chain fits on to the hook on the ceiling cap while the other end fits on to the metal loop on the lamp. Chains are made in various lengths and can, in addition, be shortened or extended by the supplier to meet your needs.
Metal loop This has two purposes: that of connecting the chain to the lamp and that of securing the nipple that protrudes through the holes in the vase caps from the inside of the lamp. A metal loop differs from a finial in that it has a hollow central passage to allow the electrical cord to pass through it to be connected to the light socket below. Various shapes and sizes are available.
Vase caps These should be selected with the same considerations of style and sizes as those for table lamps. Remember that lamps that contain crowns (as in the one described in chapter 9) cannot use vase caps and must, instead, be hung with a spider or crossbar.
Nipples These are threaded pipes that, on the one end, screw into the light socket and, on the other, into the metal loop on the top of the shade. They are made in various lengths and widths, and can, in addition, be cut with a saw when necessary.
Light sockets These, like those on table lamps, vary in shape, size and design; common varieties being made of brass, bakelite, and porcelain. The choice of whether to purchase them with or without an on-off chain or a pull-push switch is a personal one – with functionality being the determinant. Larger shades require clusters of two, three, or four light sockets to diffuse the light evenly.
Wiring a hanging lamp The cord is fed through the ceiling cap, down the chain, through the metal loop and nipple, and, finally, into the light socket. I recommend that you feed the cord through every third or so link in the chain, making certain that it

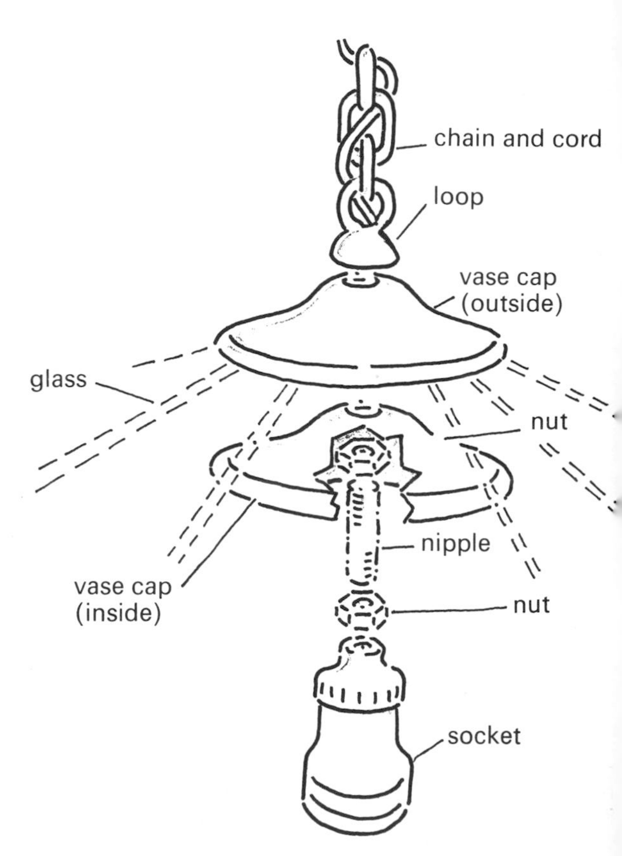

Diagrammatic representation of the fittings for a typical hanging lamp is always longer than the chain so that it is at *no* stage supporting the weight of the shade. Check, while doing this, that none of the links are enmeshed so that no sudden link slippage can occur which might break the cord or force it out of the socket.

Hanging lamps: How best to display them

Ceiling or wall lamps do, if fixed above eye-level, present a problem in that you can see inside them. Not only the bulb, but everything else that goes to make up the undercarriage of the shade – the socket, nipple, nuts, etc – will be visible; none of which are intended to be eye-catching but which often are, for exactly the wrong reasons. Not only do such electrical fittings detract from the overall beauty of the lamp, but the viewer sees the inner surface of the glass by the glare of the reflected light rays of the bulb, and not from the outside surface by transmitted light as it should be seen for maximum effect.

One sees all too many hanging lamps, that one can only assume are beautiful, hung or fixed to a wall bracket so high up that, while all of their insides can be seen, only the lower inch or so of the outside of the lamp skirt is visible. To remedy this one should either extend the length of the chain from which the lamp is hanging or turn the lamp bottom side up and fasten it directly to the ceiling or to the chain. There is, in this respect, no one correct method of hanging a lamp; the only criterion being that as much of the glass that you wish to be seen, is, in fact. Most shades, especially hemispheres or dome-shaped ones, can be at least as attractive when hung upside down as when they are 'right' side up. A glass lamp's unique attraction is in the luminosity of the glass when backlit, so the higher that a lamp is positioned above eye-level and the less of the outside of the glass that can be seen if it is hung in the conventional way, ie, with its crown upwards, then the less decorative it will be.

An alternative method of alleviating this problem, but one that I feel is less satisfactory, is to use a large frosted Duralite globe instead of a standard

An inverted hanging lampshade with a cast, metal tassle attached to the bottom. All of the glass is visible to the viewer
Reproduced by courtesy of Light Opera, Ghirardelli Square, San Francisco, California

A lampshade attached directly to the ceiling. Again, all the glass is visible

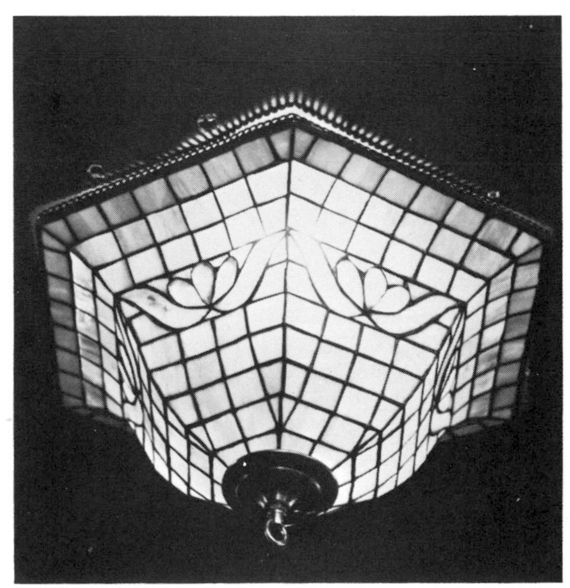

one. A globe both softens the glare of the naked bulb and, because of its large spherical shape, obscures the electrical fittings above it, thereby making the best of a bad show. Globes are made with a top metal holder that contains three thumb screws which, when screwed inwards, hold the globe in position, and which, when screwed outwards, release the globe, thereby enabling you to change the bulb when necessary. The complete globe unit is sold in various sizes that, however, require a special device which can be obtained from an electrical supply shop to link them on to standard lamp fittings. The extra effort is, though, more than compensated for by their neat appearance.

Filigree borders, collars These are very specialized items usually only stocked by stained glass suppliers and which, when used with discretion, can add that extra, sometimes indefinable, touch of quality to either a table or hanging lamp. Their over-use, however, will drastically cheapen it, as too much filigree has a decidedly kitsch look about it.

Filigreed borders and collars are stamped bands of metal, usually brass or lead, that are used either, or both, to hide joints which, because they are sharply angular, are difficult to solder neatly (the collar used in the lamp in chapter 9 was added for this reason: to hide the joint between the crown and the body of the lamp) or to be superimposed as an ornamental overlay on the glass.

These borders are pressed from very thin material and are, therefore, malleable and can be wrapped and tacked (soldered) on to the joints along the entire bottom edge of a lamp to give it an ornate finish. They are best used in shades that have large panels of a single colour as they will obviously both obscure and detract from any linear design in the glass over which they are positioned.

To attach either a border or a collar wrap them around the glass as tightly as you can and then mark the point where the end and the beginning meet. Next, carefully cut off the end with metal shears so that the two edges of the resulting joint are flush. Then rub the area of the joint very gently with a wire brush to remove any oxidation on the metal, flux it, and apply as little solder as necessary to bond the two ends together (you will best solicit an extra pair of hands for this operation to prevent the two ends from slipping). The less conspicuous the joint, the tidier it will be. Finally, add an antiquing agent to the entire filigree to give it a uniform colour, thereby further disguising the joint.

Filigree bands that form the lower border of a lamp can, in addition, be drilled or punched with holes into which various pendants can be hooked by wire. Fringes such as glass prisms, crystals, and strings of beads are some such items. Such frillery is, though, very characteristic of a particularly fussy decorative era – the late 1900s – and not readily transferable to other styles of lamps: in fact, I would recommend, while certainly being aware that such lamp accessories are available, that in general you plan to avoid them as they are of questionable taste and somewhat *passé*. The more embellishments you add to the glass the more cluttered it will appear, with addenda such as glass 'teardrop' crystals hanging from the skirt of a lamp often being more meretricious than artistic.

Wall bracket improvisation. An old English gas-lamp holder turned upside down to form a wall-bracket
Aztec lampshade by Alastair Duncan, London

Lighting considerations for lamps

As a general rule, I would recommend that the more transparent the glass in the lamp, the lower the wattage of the globe used. Conversely, the less transparent the glass, the brighter the bulb can be. You can, in addition, experiment with the wide range of bulbs – frosted and coloured – now commercially available. There are, also, numerous lighting devices, such as dimmer switches, rheostats, and other light modulators, which can be regulated from a wall-switch to produce any number of changeable colour effects and moods.

Signature

For some inexplicable reason, most people who would not hesitate to have their names emblazoned proudly across a painting would not, however, consider signing their creations in glass, and such illogicality is not lessened by the fact that it is difficult to make one's mark, so to speak, neatly and permanently on lead or glass without the proper tools. There are, though, three good reasons against anonymity in your work. First, your signature gives it a personal and professional finish by adding that extra little something to show your pride in the finished object. Secondly, and related to this, you will find that the fact that you know that you will, finally, affix your name to the end-product acts as a form of psychological quality control that will discourage you from allowing any shoddy workmanship to go uncorrected. This applies particularly to messily soldered joints that you tend to promise yourself, in all good faith, that you will return to later, but which, somehow, you invariably never quite get to do. Thirdly, you will find that if and when you decide to sell what you have made, that most buyers do, and indeed *should*, expect that it be signed. An autographed object of art is a far better potential investment to a purchaser than one that is not, which, in turn, if nothing else, often helps to narrow the gap between what you and he feels your creation is worth pricewise!

Because the glass in this lamp is transparent, so a globe is used to soften the glare of the bulb

A stamped, filigree border that can form an attractive overlay for the glass

There are basically two ways of signing a leaded glass object: either on the glass or on the leading. The former is best done with a scriber specially designed for this purpose. It has a stylus which enables you to scratch your name or monogram into the surface of the glass. Be careful, though, when using it, not to apply too much pressure or to make your signature too big because a scriber performs the same function as a glass wheel cutter (they are both made of tungsten carbide) in that it actually scores the glass, so too much force could generate a fracture.

The method of applying your signature to the leading can be done in similar fashion by carving into the surface of the lead came or soldered copper foil, but I do not recommend this as it is impermanent and looks amateurish. The oxidation of the lead or the darkening effect of an antiquing agent, if you are using one, will tend to cause the lettering to fade into its background. It can, in addition, be inadvertently or intentionally filed off the surface of the leading. No, the best procedure is to order a batch of copper or brass name tags, about $\frac{1}{8}$ in. wide with your name inscribed on them, from a sign-maker. These can be fluxed and soldered onto the leading to form a permanent and very professional finish to your projects. The conventional position for these is, on lampshades, on the inside lower border leading of the skirt, and, on windows, on the outside perimeter leading, though you can, of course, position them wherever you choose.

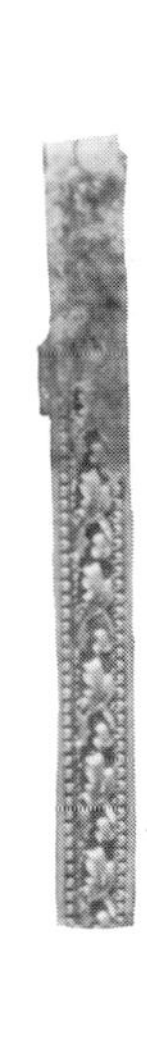

Two filigree bands of metal that can be used to form the collar between the body and the crown on a lampshade

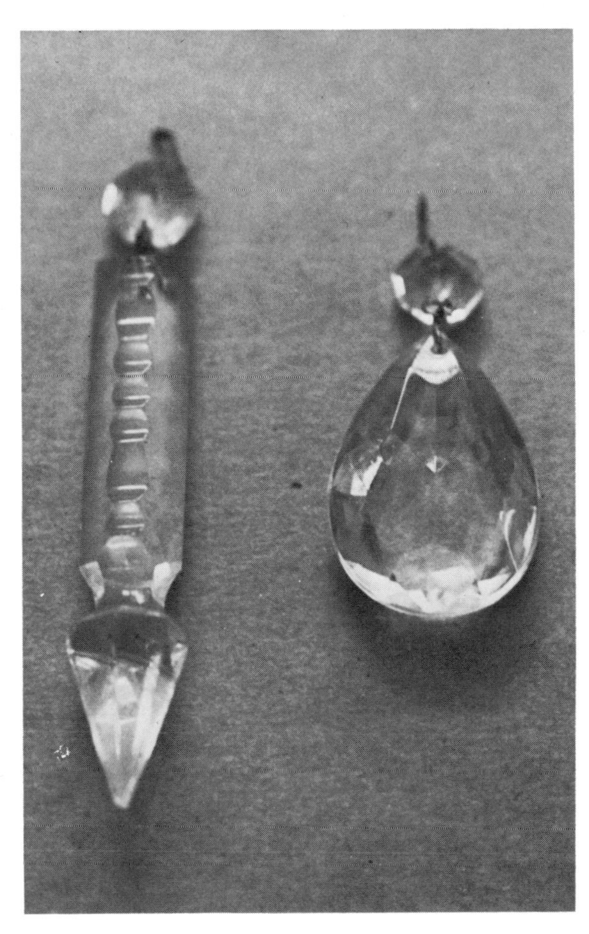

Two typical glass crystal pendants

Two brass name-tags: one unsoldered and the other soldered on to the inside border of the shade and then darkened with an antiquing solution

Tiffany's vast, two-storey, glass mosaic proscenium curtain, at the Instituto Nacional de Bellas Artes in Mexico City, depicting the twin volcanoes of Ixtacitiuatal and Popocatépetl. The curtain, which moves up and down as a frontdrop for the theatre's stage, faces a conservation problem in that the theatre (built, as is all of Mexico City, over the marshes that were once part of the Aztec capital city of Tenochtitlarn) is slowly sinking which, in turn, is threatening to crush the curtain

Chapter 13

Variations on a theme

This book has, up until now, described the methods of combining glass and lead came or copper foil to make two main glass art forms: windows and lampshades. These do, doubtlessly, constitute the vast majority of projects undertaken in leaded glass, but that this is so is due more to the fact that most people – both the public in general and, less understandably, the glass artist himself – are conditioned to thinking of glass within the limits of these two shapes. Such confinement is certainly not, however, due to any restrictions in the materials themselves. On the contrary, the remarkable versatility of copper foil in particular, and lead came to bond pieces of glass rigidly together at any angle, plus the ever increasing range of available lighting fixtures now available, allows one to create a multiplicity of both useful and beautiful glass objects. There is, in fact, almost no limit to what can be done with this art medium: multi-dimensional glass sculpture, room partitions, sign-boards, clocks, mirrors, fire screens, free-form objects, mobiles, bed valences and canopies, ceilings, coffee-tables, mosaics, terraniums . . . the list of potential uses goes on and on . . . being limited only by the extent of the artist's creativity.

The rest of this chapter illustrates just a few of the ways in which glass can, and has, been put to practical and aesthetic use. I hope it will stimulate the reader to follow suit; to start thinking of glass as a potentially functional and decorative material for all parts of the home. As a secular art form it has been neglected for far too long.

Moonrider A piece of free-form glass sculpture by Harriet Hyams, Teaneck, New Jersey. It is suspended by a stainless steel cable and has a reinforcing rod incorporated into it to provide structural support

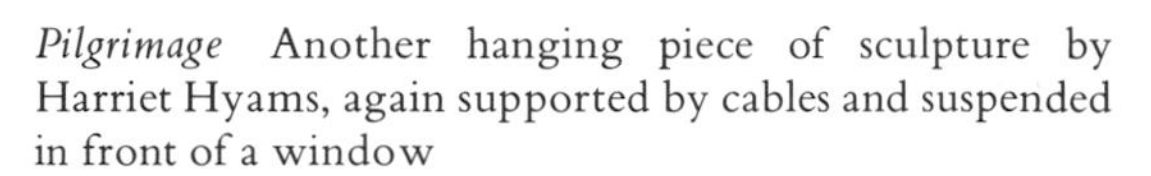

Pilgrimage Another hanging piece of sculpture by Harriet Hyams, again supported by cables and suspended in front of a window

Sign-boards, either as name plates for the home or as advertising for a shop, are most effectively made by means of a thin, rectangular box. The sides consist of the design made up of opalescent glass, which are illuminated at night by tubes of strip lighting placed inside the box

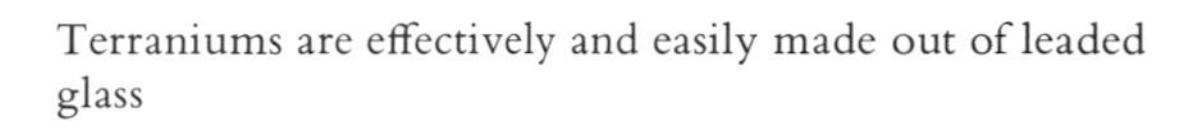

Terraniums are effectively and easily made out of leaded glass

A telephone booth incorporating leaded glass
Sal Fiorito, Washington DC

A three-piece screen of mahogany and leaded glass used as a room divider
Tom Rodriquez, San Rafael, California

Epilogue

"That's the reason they're called lessons,"
the Gryphon remarked, "because they lessen
from day to day."
Alice in Wonderland

We have come to the end of what I hope has been both edifying and enjoyable. The first part of the book set out to explain the various tools and materials used in leaded glass work and the method of applying them. The second part – while offering several ideas on how to implement the techniques explained in the first section – has, though, offered less and less new in each succeeding chapter; only variations on a theme, in fact. Once you have learnt the method of working with glass and lead or copper foil then the way in which you utilize this information is up to you. The introductory projects explained in this book should only – if I have achieved what I set out to do – whet your creative appetite to invent your own shapes and designs.

May I, finally, make an appeal on behalf of all crafts. If you are looking for a hobby somewhere on terra firma between parachuting and potholing, or if you experience a sense of taedium vitae – then I recommend that you try your hand at a craft; whether glass, pottery, weaving, or whatever. The craft industry in England needs a continual influx of new recruits if it is to remain an integral part of English heritage. So if you find that there are moments when you are in need of something to do, then recall Rudyard Kipling's:

The Camel's Hump is an ugly lump
which well you may see at the zoo;
But uglier yet is the Hump we get
from having too little to do.

or Benjamin Franklin's:

Dost thou love life? Then do not
squander time, for that's the stuff
life's made of

. . . and take up a craft. The rewards are manifold.

General subject reading

ARMITAGE, E LIDDELL *Stained Glass* Branford, Newton Centre, Mass, 1959

KOCH, ROBERT *Louis C Tiffany, Rebel in glass* Crown, New York, 1964: *Louis C Tiffany's Glass – Bronzes – Lamps* Crown, New York, 1971

NEUSTADT, DR EGON *The Lamps of Tiffany* The Fairfield Press, New York, 1970

PIPER, JOHN *Stained Glass: Art or Anti-Art* Studio Vista, London: Van Nostrand Reinhold, New York, 1968 (out of print)

SOWERS, ROBERT *The Lost Art* Lund Humphries, London, 1954 (out of print): *Stained Glass: an architectural Art* A Zwemmer, London, 1965 (out of print)

REVI, ALBERT C *Art Nouveau Glass* T Nelson & Sons, Camden, New Jersey, 1968

Technical books

DIVINE, J A F and BLACKFORD, G *Stained Glass Craft* Dover, New York, 1972

DUVAL, JEAN-JACQUES *Working with Stained Glass* Cowell, New York, 1972

ERIKSON, ERIK *Step-by-Step Stained Glass* The Golden Press, New York 1974

HAMILTON, WALTER J *The Technique of Making Leaded Glass Ornaments* Hamilton Studios, 2201 Musgrove Road, Maryland, 20904, 1971: *The Advanced Techniques of Making Leaded Glass Projects*, 1972: *Stained Glass Patterns by your request*, 1974

HEPBURN, JAMES H *Creating Stained Glass Lampshades* President Press, 1974

ISENBERG, ANITA and SEYMOUR *How to Work in Stained Glass* Chilton Books Co, Philadelphia, 1972: *Stained Glass Lamps* Chilton Book Co, Radnor, Pa, 1974

LEE, LAWRENCE *Stained Glass* Oxford University Press, London, 1967

LUCIANO *Stained Glass Lamps* Hidden House Press, Palo Alto, California, 1973

MOLLICA, PETER *Stained Glass Primer* Mollica's, 1940A Bonita, Berkeley, California 94704, 1973

NERVO, JOANNE *Stained Glass Patterns* Nervo Studios, 2027 7th Street, Berkeley, California 94704, 1973, 1974

REYNTIENS, PATRICK *The Technique of Stained Glass* Batsford, London, 1967; Watson-Guptill, New York, 1967

STAINED GLASS CLUB *The Glass Workshop Bimonthly Magazine* PO Box 244, Norwood, New Jersey, 07648

Suppliers UK

Glass, glass jewels, soldering irons, glass cutters, copper foil tape, U-frame leads, moulds, patina
James Hetley and Co Ltd
10 Beresford Avenue
Wembley HAO 1RP
Middlesex

H-frame leads
Lonsdale Metal Co
608 High Road
London E10

U-frame leads
Brunton & Co Ltd
5 Miles Road
Vauxhall
London SW8

Lamp accessories
See Yellow Pages under *Lighting Fixtures, Supplies & Parts*
Recommended for London:
W Sitch & Co
48 Berwick Street
London W1

Glazing tools
Berlyne, Bailey & Co Ltd
29 Smedley Lane
Cheetham
Manchester M8 8XB
Emile Regniers & Co
450 High Road
Ilford
Essex
Buck & Ryan Ltd
101 Tottenham Court Road
London W1

Manufacturers

Soldering irons
Weller Electric Co
Redkiln Way
Horsham
Sussex

Suppliers USA

Complete stockists
Nervo Studios
2027 7th Street
Berkeley
California 94710
Whittemore-Durgin Glass Co
Box 2065
Massachusetts 02339
Glass Masters Guild,
52 Carmine Street,
New York, N.Y. 10014

Miscellaneous
Glass Workshop
PO Box 244
Norwood
New Jersey 07648

Marbleized mirror
Felio Pereantón, SA
Plaza de Vazquez Mella 5
Madrid
Spain

Patterned moulds
H L Worden Co
Box 519
Granger
Washington 98932